Sustainable Development and Biodiversity

Volume 18

Series editor

Kishan Gopal Ramawat
Botany Department, M.L. Sukhadia University, Udaipur, Rajasthan, India

This book series provides complete, comprehensive and broad subject based reviews about existing biodiversity of different habitats and conservation strategies in the framework of different technologies, ecosystem diversity, and genetic diversity. The ways by which these resources are used with sustainable management and replenishment are also dealt with. The topics of interest include but are not restricted only to sustainable development of various ecosystems and conservation of hotspots, traditional methods and role of local people, threatened and endangered species, global climate change and effect on biodiversity, invasive species, impact of various activities on biodiversity, biodiversity conservation in sustaining livelihoods and reducing poverty, and technologies available and required. The books in this series will be useful to botanists, environmentalists, marine biologists, policy makers, conservationists, and NGOs working for environment protection.

More information about this series at http://www.springer.com/series/11920

Dilip Nandwani
Editor

Urban Horticulture

Sustainability for the Future

Springer

Editor
Dilip Nandwani
Department of Agriculture
and Environmental Sciences,
College of Agriculture
Tennessee State University
Nashville, TN
USA

ISSN 2352-474X ISSN 2352-4758 (electronic)
Sustainable Development and Biodiversity
ISBN 978-3-030-09784-4 ISBN 978-3-319-67017-1 (eBook)
https://doi.org/10.1007/978-3-319-67017-1

Printed on acid-free paper

This Springer imprint is published by the registered company Springer International Publishing AG part of Springer Nature
The registered company address is: Gewerbestrasse 11, 6330 Cham, Switzerland

Foreword

Bringing Innovation in Urban Agriculture in the Twenty-first Century

In a rapidly urbanizing world, urban agriculture represents an opportunity for improving food supply, health conditions, local economy, social integration and environmental sustainability altogether. Accordingly, while a diversity of farming systems is encountered in the different world regions, it is estimated that about a third of urban dwellers is involved worldwide in the agro-food sector. The current scarcity of knowledge on urban agriculture has somehow hindered the relevance of this activity, although several researches have recently spent a great effort in addressing main drawbacks and opportunities of this growing sector.

The mentioned interest on urban agriculture has led, in the recent decades, to the sprouting of a diversity of projects, both guided and promoted by governments than born by bottom-up community-based initiatives. At city scale, the distribution of such projects can lead to the creation of green corridors that have been also referred to as 'continuous urban productive landscapes', whose main benefits also include permeability to biodiversity and functionality towards climate change adaptation and resilience. Accordingly, the concept of edible urban landscapes (edible cities, foodscapes) is today finding application all over the world.

However, when plant cultivation gets inside cities, a first limit is set by space availability. Hence, the implementation of innovative growing systems generally occurs through adaptation of state-of-the-art technologies developed in the rural contexts and/or by integration of growing solutions specifically designed for the urban concrete infrastructures. These are the cases of the building integrated rooftop greenhouse, the vertical cultivation of plants on greened building facades or even the conversion of basements or abandoned building into plant factories through artificial lighting using LED technologies.

Furthermore, right after space availability, limited light access due to reciprocal shading from buildings often results in plant cultivation being unfeasible. That is where the above-mentioned technologies come in, either moving up the plants until where the light becomes sufficient (e.g. greened roofs or walls) or providing functional light in closed urban environment (e.g. indoor plant factories).

Once plant cultivation within the urban fabric becomes technically feasible, it is crucial to define its sustainability, addressing how its three spheres (social, economic and environmental) are affected by a single or a combination of projects.

From a social perspective, urban agriculture is often a mean for the promotion of intercultural dialogue, social inclusion and community building. Urban gardens are also found in schools, where they offer optimal teaching tools for different disciplines, also through application of experiential learning methods.

On the other hand, assessing the economic sustainability of urban agriculture is often a quite complex and eventually controversial exercise. This since, beyond food production, the quantification and monetization of the associated benefits need to be appropriately addressed. Recent researches have shed some light onto the financial and economic evaluation of urban agriculture projects, particularly addressing the definition of viable business model typologies.

Finally, from an environmental perspective, urban agriculture can directly reduce the city footprint since it reduces the impact associated with food transport, but also by improving the resource efficiency of the urban ecosystem (e.g. when organic wastes are composted, or rainwater is collected), or mitigating the urban heat island effect through plant transpiration, resulting in a reduction of the city emissions of carbon dioxide.

As it appears, there is an urgent need for defining the multifaceted aspects of urban agriculture through innovative and interdisciplinary studies. The present book provides a comprehensive overview of the sector with a specific focus on how to bring sustainable solutions for plant cultivation in cities.

The editor, Dr. Dilip Nandwani, possesses a research and extension background that ranges from organic vertical systems, vegetable and medicinal crop production and improvement, plant tissue culture and propagation, but also multifaceted social dimensions, such as the contribution that sustainable agriculture can play in food security, nutrition, income generation and reduction of gender inequalities in small-scale farming systems. This interdisciplinary background enabled him to wisely assemble and edit this book.

Bologna, Italy

Francesco Orsini
Research Centre on Urban Enviornment for
Agriculture and Biodiversity (RESCUE-AB)
Department of Agricultural Sciences, Alma Mater
Studiorum—Bologna University

Acknowledgements

This is my third book in the book series 'Biodiversity for Sustainable Development' and like previous two books, this was not an easy task to complete without the support of and valuable suggestions and encouragement received from Prof. K. G. Ramawat (series editor), contributions from the authors, students, colleagues and Dean, College of Agriculture, Tennessee State University. My sincere appreciation to Springer publication team working with me diligently throughout the project. Special thanks to my family members, wife Varsha, children (Gayatri & Rahul) and parents, their love and patience enabled me to complete this project successfully.

Nashville, USA
2017

Dilip Nandwani, Ph.D., CPH

Contents

Editor and Contributors

About the Editor

Dr. Dilip Nandwani Associate Professor with MS and Ph.D. from the University of Jodhpur (now Jai Narayan Vyas University), Jodhpur, India. He joined the Tennessee State University as Research Associate Professor (Organic Agriculture) in 2014. Prior to joining Tennessee State University, he served University of the Virgin Islands as Research Associate Professor (Horticulture) and spent a decade in the American Pacific. He has been working with Land Grant institutions for almost two decades in agricultural research, extension and teaching. He has widely published in the area of production agriculture and horticultural sciences. His research and extension work has lead over 130 articles (peer-reviewed, extension booklets and conference proceedings). He is editor of three books and Certified Professional Horticulturist (CPH) from the American Society for Horticultural Science (ASHS). He worked with United Nations Environment Programme (UNEP-GEF) as Regional Advisor. Dr. Nandwani teaches Principles of Organic Agriculture and Organic Certification courses, received several awards, obtained over $2.6M grants from regional, national and international organizations, consulted and travelled globally.

Contributors

K. G. Arvanitis Department of Natural Resources Management and Agricultural Engineering, Agricultural University of Athens, Athens, Greece

S. K. Basu Department of Biological Sciences, University of Lethbridge, Alberta, Canada

L. Battisti Department of Agriculture, Forest and Food Sciences, University of Turin, Turin, Grugliasco, Italy

Giovanni Bazzocchi Research Center for the Urban Environment on Agriculture and Biodiversity (RESCUE-AB), Department of Agricultural Sciences (DipSA), University of Bologna, Bologna, Italy

L. Bianco TEBE Research Group, Energy Department, Politecnico di Torino, Turin, Italy

W. Cetzal-Ix Instituto Tecnológico de Chiná, Campeche, Mexico

Eva Cuerva Department of Construction Engineering, School of Industrial Engineering (ETSEIB), Universitat Politècnica de Catalunya (UPC-Barcelona Tech), Barcelona, Spain

Mireia Ercilla-Montserrat Sostenipra Research Group (SGR 01412), Institute of Environmental Sciences and Technology (MDM-2015-0552), Universitat Autònoma de Barcelona (UAB), Bellaterra, Barcelona, Spain

K. P. Ferentinos Department of Agricultural Engineering, Institute of Soil and Water Resources, Hellenic Agricultural Organization "Demeter", Athens, Greece

Mohammed Ibrahim Fetouh Department of Horticulture, Faculty of Agriculture, Tanta University, Tanta, Egypt

Xavier Gabarrell Department of Chemical, Biological and Environmental Engineering, Institute of Environmental Sciences and Technology (MDM-2015-0552), Sostenipra Research Group (SGR 01412), Universitat Autònoma de Barcelona (UAB), Bellaterra, Barcelona, Spain

Giorgio Gianquinto Research Center for the Urban Environment on Agriculture and Biodiversity (RESCUE-AB), Department of Agricultural Sciences (DipSA), University of Bologna, Bologna, Italy

R. Giordano Department of Architecture and Design, Politecnico di Torino, Turin, Italy

John Ikerd Agricultural and Applied Economics, Division of Applied Social Sciences, University of Missouri, Columbia, MO, USA

Alejandro Josa Department of Civil and Environmental Engineering, School of Civil Engineering (ETSECCPB), Institute for Sustainability Science and Technology (IS.UPC), Universitat Politècnica de Catalunya (UPC-BarcelonaTech), Barcelona, Spain

R. N. Kulloli Botanical Survey of India, Arid Zone Regional Circle, Central Arid Zone Research Institute, Jodhpur, Rajasthan, India

Suresh Kumar Central Arid Zone Research Institute, Jodhpur, Rajasthan, India; Central Arid Zone Research Institute, Jodhpur, India

F. Larcher Department of Agriculture, Forest and Food Sciences, University of Turin, Turin, Grugliasco, Italy

Pere Llorach-Massana Sostenipra Research Group (SGR 01412), Institute of Environmental Sciences and Technology (MDM-2015-0552), Universitat Autònoma de Barcelona (UAB), Bellaterra, Barcelona, Spain

Francesca Magrefi STePS s.r.l, Treviso, Italy

Nicola Michelon Research Center for the Urban Environment on Agriculture and Biodiversity (RESCUE-AB), Department of Agricultural Sciences (DipSA), University of Bologna, Bologna, Italy

E. Montacchini Department of Architecture and Design, Politecnico di Torino, Turin, Italy

Juan Ignacio Montero Institute of Food and Agricultural Research (IRTA), Barcelona, Spain

Pere Muñoz Institute of Food and Agricultural Research (IRTA), Barcelona, Spain

Ana Nadal Sostenipra Research Group (SGR 01412), Institute of Environmental Sciences and Technology (MDM-2015-0552), Universitat Autònoma de Barcelona (UAB), Bellaterra, Barcelona, Spain

Dilip Nandwani Department of Agriculture and Environmental Sciences, College of Agriculture, Tennessee State University, Nashville, TN, USA

E. Noguera-Savelli Catedrática CONACYT, Colegio de Postgraduados Campus Campeche, Sichochac, Champotón, Campeche, Mexico

D. G. Noh-Contreras Instituto Tecnológico de Chiná, Campeche, Mexico

Sochinwechi Nwosisi Department of Agriculture and Environmental Sciences, College of Agriculture, Human and Natural Sciences, Tennessee State University, Nashville, TN, USA

Francesco Orsini Research Center for the Urban Environment on Agriculture and Biodiversity (RESCUE-AB), Department of Agricultural Sciences (DipSA), University of Bologna, Bologna, Italy

P. Papageorgas Department of Electrical and Electronics Engineering, University of West Attica, Egaleo, Athens, Greece

Giuseppina Pennisi Research Center for the Urban Environment on Agriculture and Biodiversity (RESCUE-AB), Department of Agricultural Sciences (DipSA), University of Bologna, Bologna, Italy

Anna Petit-Boix Sostenipra Research Group (SGR 01412), Institute of Environmental Sciences and Technology (MDM-2015-0552), Universitat Autònoma de Barcelona (UAB), Bellaterra, Barcelona, Spain

Oriol Pons Department of Architectural Technology, School of Architecture (ETSAB), Universitat Politècnica de Catalunya (UPC-Barcelona Tech), Barcelona, Spain

W. S. Poot-Pool Instituto Tecnológico de Chiná, Campeche, Mexico

D. Piromalis Department of Industrial Design and Production Engineering, University of West Attica, Egaleo, Athens, Greece

Joan Rieradevall Sostenipra Research Group (SGR 01412), Institute of Environmental Sciences and Technology (MDM-2015-0552), Universitat Autònoma de Barcelona (UAB), Bellaterra, Barcelona, Spain; Department of Chemical, Bilogical and Environmental Engineering, Universitat Autònoma de Barcelona (UAB), Bellaterra, Barcelona, Spain

Mary A. Rogers Sustainable & Organic Horticultural Food Production Systems, Department of Horticultural Science, University of Minnesota, Saint Paul, MN, USA

Maria Rosa Rovira Sostenipra Research Group (SGR 01412), Institute of Environmental Sciences and Technology (MDM-2015-0552), Universitat Autònoma de Barcelona (UAB), Bellaterra, Barcelona, Spain

David Sanjuan-Delmás Sostenipra Research Group (SGR 01412), Institute of Environmental Sciences and Technology (MDM-2015-0552), Universitat Autònoma de Barcelona (UAB), Bellaterra, Barcelona, Spain

Esther Sanyé-Mengual Department of Agricultural Sciences (DipSA), Research Center for the Urban Environment on Agriculture and Biodiversity (RESCUE-AB), Alma Mater Studiorum, University of Bologna, Bologna, Italy

V. Serra TEBE Research Group, Energy Department, Politecnico di Torino, Turin, Italy

Monique Centrone Stefani Department of Sociology and Social Research, University of Trento, Trento, Italy

S. Tedesco Department of Architecture and Design, Politecnico di Torino, Turin, Italy

Abbreviations

AMS	Agricultural Marketing Service
ANCeSCAO	National Association of Social Centres, Committees and Gardens of the Elders
CBD	Convention on Biological Diversity
CITES	Convention on International Trade in Endangered Species of Wild Fauna and Flora
CSA	Community Supported Agriculture
CTE	Technical Code of Edification
FAO	Food and Agriculture Organization of the United Nations
FAO	Food and Agriculture Organization
GIS	Geographical Information Systems
ICP	Catalan Institute of Palaeontology Miquel Crusafont
ICP-OES	Inductively Coupled Plasma Optical Emission Spectrometry
ICTA	Institute of Environmental Science and Technology
i-RTG	Integrated Rooftop Greenhouse
ISO	International Standard Organization
LCA	Life Cycle Assessment
LCC	Life Cycle Cost
LDPE	Low-density Polyethylene
LIDAR	Light Detection and Ranging
LWIR	Long Wave Infrared
LWS	Living Wall Systems
NHF	New Haven Farms
RTG	Rooftop Greenhouse
SCT	Social Cognitive Theory
UA	Urban Agriculture
UHI	Urban Heat Island
UPE	Urban political ecology
USDA	United States Department of Agriculture
VF	Vertical Farming

Part I
Urban Horticulture Tools and Techniques

Chapter 1
Urban Horticulture: Overview of Recent Developments

Sochinwechi Nwosisi and Dilip Nandwani

Abstract The growing populations in urban areas and the decrease in available land are major constraints to sustainable crop production in and around cities. All over the world, many people do not get enough food and proper nourishment. Urban horticulture which is the production of vegetables, herbs, medicinal and ornamental plants around cities may serve as a solution. Worldwide, temperatures vary in different seasons and urban growers face challenges in sustaining production due to poor yield and crop losses. Therefore, urban horticultural growers may need to employ new techniques and tools to improve their practices. However, many of these recent developments require capital, access to specific information, knowledge, and resources to improve their skills on production, processing, and marketing of their produce. This paper reviews the recent developments in urbanization and urban horticulture, the types of urban horticulture as well as tools and techniques in urban horticulture. The focus of this paper is on recent developments in vertical farming systems (hydroponics, aeroponics, and aquaponics) including organic vertical farming research. Also listed in this study are solutions that urban horticulture provides such as provision of food security and, poverty alleviation. Some of the concerns, challenges, and benefits in urban vertical farming systems would also be explored.

Keywords Urban agriculture · Food security · Water efficiency · Land conservation · Sustainable farming

1.1 What Is Urban Horticulture?

Each time climatic temperature increases by 1°, it is evaluated that 10% of the land presently under crop development will become unavailable. It is possible that there

D. Nandwani (✉)
Agriculture and Environmental Sciences, College of Agriculture, Tennessee State University, Nashville, TN 37209, USA
e-mail: dnandwan@tnstate.edu

S. Nwosisi
Agricultural and Environmental Science, College of Agriculture, Human and Natural Sciences, Tennessee State University, Nashville, TN 37209, USA

D. Nandwani (ed.), *Urban Horticulture*, Sustainable Development and Biodiversity 18, https://doi.org/10.1007/978-3-319-67017-1_1

will be insufficient land in the near future, suitable for farming that would provide food for the total world populace (Anonymous 2011). It has also been discovered that approximately 80% of the land currently under cultivation worldwide is being used (Anonymous 2011). A large percentage of the land on which the world's sustenance is built has turned out to be depleted or no longer usable (Fabian and Malte 2011). Urban horticulture described, as the production, preparation, and supply of food products and other materials through the meticulous growth of plants and rearing of animals within and around urban areas, can help solve this challenge.

The definition of urban and horticulture varies with country, local culture, and traditions (China Development Research Foundation 2010). Horticulture is simply defined as the establishment of gardens or cultivation of plants while agronomy, on the other hand, is the growth of forests, trees, and related products, cereals or animal fodder or simply agriculture-farming practices (Albert 2012). Horticulture is that part of agriculture that involves growing food for food, medicine, or beauty. Agronomy or field cropping, on the other hand, majorly involves the cultivation of nonperishable crops such as leguminous crops, oilseeds, and fiber (Fig. 1.1).

Urban horticulture may involve cultivation of ornamental and medicinal species such as fruit, vegetables, tubers, herbs, aromatic plants, roots, mushrooms, and so on (Francesco et al. 2013) mainly in the city or its surrounding regions where there is competition for land (Moustier 1999). It could involve the use of plants for recreation, the design of landscape, public facilities, or therapy of ill and disabled individuals.

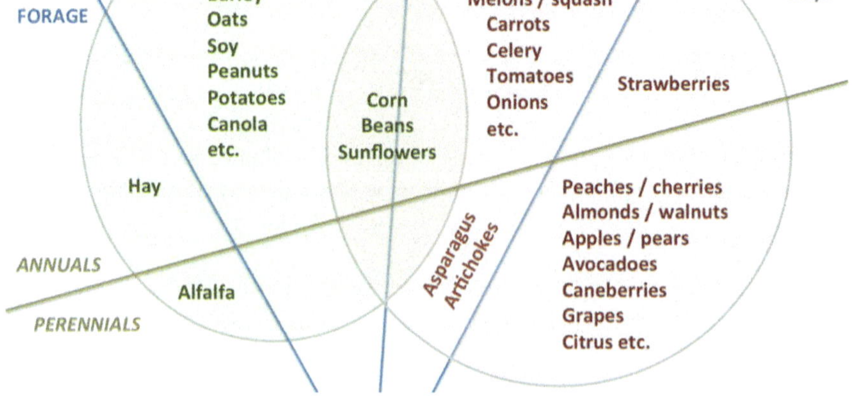

Fig. 1.1 Various ways to group crops produced in the U.S.A. *Source* Kerr (2016)

Although urban agriculture consists of plant production systems and livestock activities, horticulture or plant production is most often the main aspect.

The worth of horticultural production in the U.S. has been steady over the previous decade while the production of fruits and tree nuts has expanded consistently (Fig. 1.2). The most fruit sales have been recorded for oranges, apples, strawberries, and grapes while the primary vegetables are potatoes and tomatoes. The worth of nuts and trees has risen essentially to historical levels of about $10 billion of late, with the value of most significant tree nut crops spearheaded by pistachios, walnuts, and almonds attaining record highs (Anonymous 2016a).

There are different types of urban horticulture (Fig. 1.3). On the basis of water and soil being available, urban horticulture may be grouped into intensive horticulture making use of advanced technologies and plant nurseries on the small scale (where systems without soil and localized irrigations are used) (Drescher and Laquinta 1999); community gardens (here land is made available by the local government and such land is further divided into plots of 100 m each) (Francesco et al. 2013); and micro- or mini gardens which also use simple soilless systems to grow flowers, herbs, vegetables, mushrooms, etc.

In developed nations, urban horticulture is frequently practiced on small areas of leased land, privately owned land or on community gardens such as school gardens in city corners (Van Leeuwen et al. 2010). Apart from the advantage of food provision for home consumption and other recreational benefits, urban horticulture in these areas likewise offers an outlet for social association and cultivation of locally grown healthy foods (Nugent 2000). Conversely, in developing nations, urban horticulture is employed to generate independent businesses and a source of direct income or funds thereby prompting social security (Van Leeuwen et al. 2010). Urban farming additionally enhances access to garden-fresh nourishing foods, helps battle dia-

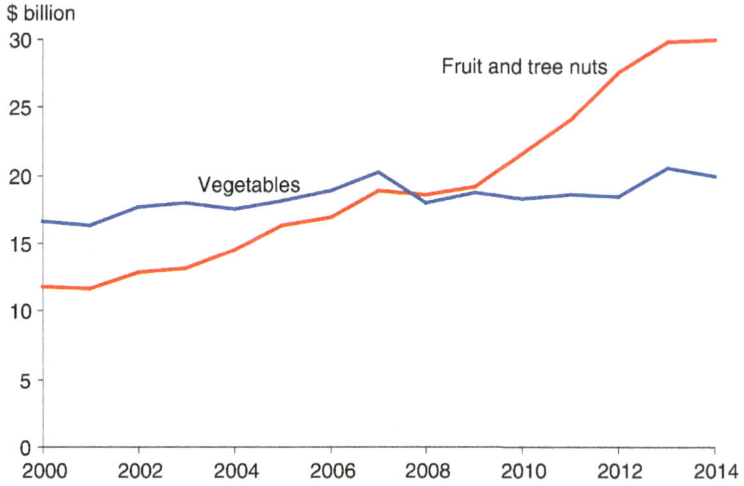

Fig. 1.2 The value of horticultural production in the U.S.A from 2000–2014. *Source* Anonymous (2016a)

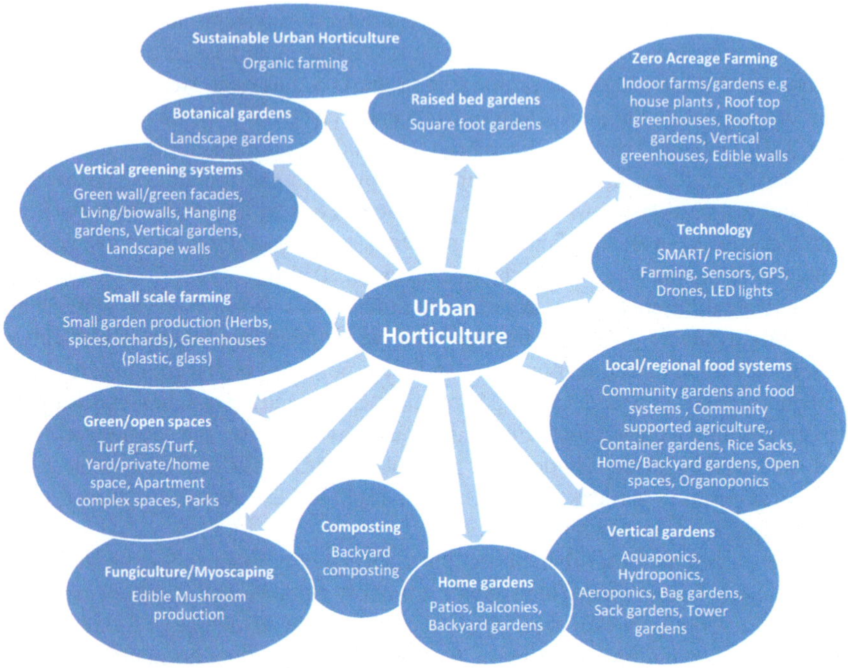

Fig. 1.3 Types of urban horticulture

betes and weight problems in children, gives accessibility to foods that are scarce among migrant communities, and improves poor health due to malnourishment in many urban areas and social well-being through strengthening existing relationships between different races and reducing criminality (Lovell 2010).

There has been an expansion in the number of individuals purchasing locally grown food products over the past few years (Yue et al. 2011). Other than its characteristics of freshness and safety, one of the reasons why consumers decide to buy food grown locally is because they believe they are supporting the economy (Yue and Tong 2009).

In addition, urban horticulture provides better livelihoods and additional income to all major stakeholders involved in the horticulture value chain (Francesco et al. 2013) and helps create economic employment in farming, marketing, and processing activities. The amount of money needed and the impact of growing food on the environment is lessened because growing food locally requires a smaller number of intermediaries, less transport, storage of perishable foods, packaging and handling (FAO 2008). Urban horticulture would help also to reduce food deserts in minority and low-income communities and help provide food in the local communities at unaffordable prices (Eva 2010). Recreational services, maintained landscapes and biodiversity would be provided by urban horticulture thereby generating a higher quality of livelihood for persons residing in urban communities (Van Veenhuizen and Danso 2007). Through urban horticulture, waste in urban areas can be managed more effectively (use of animal manure and household waste products), thereby contributing significantly to the reduction of several diseases. Lastly, urban horticulture would help promote gender equality and empower women as they are more predisposed to discrimination in urban territories and have less access to traditional salary-generating enterprises than men.

Urban horticulture is however not without challenges some of which include: (1) difficulty for poor people to get access to land in poor communities that may be solved by provision of loans with low interest and long-term land leases if land is adequately maintained, (2) lack of the required knowledge or skills by urban horticulturists to grow their plants is a challenge. Therefore, there is a need to provide affordable and accessible library of resources and network of contacts wherein growers can obtain the necessary skills, training and support, (3) challenges due to adverse weather conditions affect crops as most horticulturists try to produce food in and out of season, therefore, information from extension agencies and facility support such as greenhouse provisions could be made available, and (4) urban horticultural practices can be expensive and difficult for growers to afford and as such, loans, insurance, grants, and funding support by government, foundations, local and international agencies to support low-income families, establishment of community gardens, material/equipment donations, etc., should be provided.

1.2 Urbanization and Urban Communities

Globally, a larger number of individuals live in urban ranges than in rural communities, with 54% of the total populace dwelling in urban zones in 2014 (Fig. 1.4). Over the last three decades, the growth of urban populations has rapidly developed with greater economic growth. In 1950, 30% the total populace was urban, and by 2050, 66% of the total populace is anticipated to be urban. As the population of the world is expected to exceed nine billion by the year 2050, the part that urban horticulture would play in food security worldwide has become a major subject matter for dialogue.

Urbanization which arises due to more people in a nation living in the urban areas due to migration from the rural communities (which then experience a declining population) is leading to an expansion of the use of land (Satterthwaite et al. 2010). The effect of urbanization on agriculture would lead to a major change in how food is marketed and distributed (Kennedy et al. 2004). Urbanization has resulted in changes in diet and gender roles in the need to bring home two salaries in urban communities (Popkin 1999). Increasingly in urban areas, there is a shift toward increased consumption of fresh fruits, vegetables, meat, and dairy with improved urban incomes (Agnes 2014). Urbanization is also expected to lead to more people working in food processing, transportation, and retail industries and fewer people in agriculture (Cohen and Garrett 2009). Thus, individuals living in urban communities would need to sell specialty or highly valued crops either for consumption or as a source of income on urban agriculture (Smit et al. 1996; Redwood 2009).

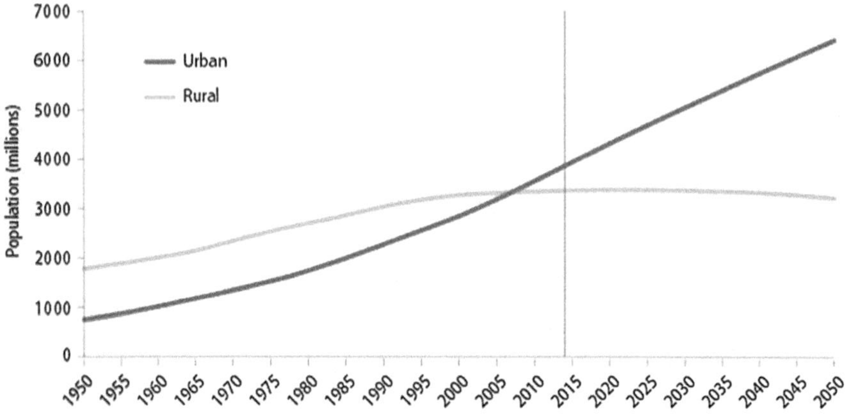

Fig. 1.4 World urban and rural population, 1950–2050. *Source* Anonymous (2014)

Several studies from East Africa in the 1990s depicted 17–36% of the population of people in urban areas in those regions grew their own crops or kept livestock (Lee-Smith 2010) and the diversity among urban farmers there consisted of students, unemployed, government officials, casual laborers, teachers, urban planners, and people who worked part-time (Sawio 1994). Health, nutrition, fresh, local and organic food advocates, faith communities, university extension services, community gardeners, emergency food distributors, environmentalists, economic developers, and other individuals, organizations, and communities have and currently are joining in the efforts to grow food in and around city spaces (Katherine et al. 2002). Most of this is a bid to assist urban dwellers who live in poverty and food insecurity and are constantly faced with hunger, anxiety of not having enough food, and lack adequate nutrition (Katherine et al. 2002).

80% of the people in the United States today reside in urban areas and with the growth in the urban population of the U.S. and that of the rest of the world comes the problem of what way is the best to feed people who do not stay in close proximity to areas where the food is actually grown (Katherine et al. 2002). There are an unacceptable number of people in the U.S. (wherein food produce is affordable for the most part, not exposed to danger or risk and is nourishing) and all around the world not having enough to eat presently. Therefore, because current systems need a lot of resources to move food from the location where it is grown to customers and supermarkets within urban areas, there exists a need to support, develop, and grow urban agriculture.

In some economies like Africa, urbanization has encouraged subsistence agriculture resulting in lack of markets and weak consumer demand as individuals attempt to grow their own produce instead of buying. In places like Asia, however, urbanization had led to the markets tilting toward a more commercial agriculture in the increasingly global markets which has been followed by quick improvements in the rates of poverty present in those areas (Ravallion et al. 2007). Still today, malnutrition is an issue in many urban communities worldwide which is more as a result of little or no income than an inability to grow food (Satterthwaite et al. 2010).

1.3 Tools and Techniques of Urban Horticulture

Although, horticulture in urban areas depends on the specific environment, challenges, opportunities, culture, and cultural practices, however, horticultural growers worldwide strive to grow crops all year round for personal consumption or the commercial market. Thus, certain tools and techniques may be required.

1.3.1 Tools of Urban Horticulture

Use of plastic covers and greenhouses. By building technical knowledge on plastic covers, greenhouses, seasonal changes in temperature, and the application of hydroponic techniques (Table 1.1), insect/pest problems in early plant growth stages may be avoided.

Irrigation and water supply. For vegetables to obtain good quality in most cases, they need to be watered twice a day. Water can be applied overhead or brought directly to the plant, using containers manually or by machine sprinklers. The manual type, though efficient and can give enough water the crop needs, is labor intensive. The drip irrigation is another method of water supply which conserves moisture by about 20% in contrast to the sprinkler system, however, it is expensive, and requires special knowledge and water to be free from dirt so as to stay clear of problems that might arise from the emitters being obstructed (Philippe and Hubert 2006). The underground method provides water to the crop by capillary action reducing the amount of pathogens in the vegetable crops by filtration of soil but installation and operation are quite difficult (Philippe and Hubert 2006).

Fertilization. Plants (especially vegetable and ornamentals) need several tons/ha/year of macro (nitrogen, phosphorus, potassium, etc.) and micro (boron, manganese, copper, etc.) elements to grow and develop good quality in freshness and tenderness (Philippe and Hubert 2006). Organic fertilizers give a good amount of micronutrients to the plant and assist in making soil structure better. However, employing the use of organic fertilizers such as using livestock and poultry manure or wastes from household/human activities or sewage sludge may lead to the pollution of soils and water in cities with heavy metals and organic waste products. Inorganic fertilizers, on the other hand, are easier to add and allow the addition of nutrients in adequate doses, however, there is the possibility of applying too much and polluting water and soils in the city with phosphates and nitrates.

Pesticides. Chemical pesticides have helped to increase yields of crops and combat pests but not without reoccurring damage to the image of the agricultural industry. Use of pesticides has also led to health problems/risks of farm workers, consumers,

Table 1.1 Tools and techniques in urban horticulture

1.	Infrastructure	Farm infrastructural facilities (storage/packaging facilities, greenhouses, irrigation, and high tunnels) and equipment (for preparation and transportation of mass produce), office space, etc., would help improve correspondence amongst suppliers and end clients. Financial and other support, for example in the United States, can be provided through irrigation specialist, university extension agents, local national resources conservation service (NRCS), rural energy for America program (REAP), sustainable agriculture research and education (SARE), and farm service agency (FSA)
2.	Access to land	Many farmers in cities find it hard to get and pay for the suitable land. In many countries around the world, development agencies in cities are working to change this and government policies are assisting in identifying available land plot for lease to growers. However, more can be done
3.	Water and soil quality	In many cities, this can be a challenge, so growers are advised to run tests to avoid contamination of their produce due to pollution from industries, sewers, land waste disposals, etc. The government environmental conservation and protection agencies, as well as university extensions and research laboratories, are good go-to resources in such situations. In addition, such agencies can help with financial and technical assistance on water use as water is a limited but essential resource in urban horticulture, therefore, helps to reduce costs of operation on water used in an efficient manner
4.	Capital and financing	Urban horticulture businesses require a significant amount of money to start up in most cases and therefore growers may require large sums of money in form of loans or grants. Farmers are encouraged to draft a good convincing business plan before sourcing manuals for potential lenders or creditors
5.	Marketing	Growers should, before planting, have and know their market (vegetables spoil quickly) and inform consumers that they are selling. Next, they should develop a marketing plan based on sound business principles and decide what marketing channel best suits their consumers' needs. Organic growers should evaluate the market region demand for organic products and they can look to certify their crop as organic in order to improve marketability. Organic growers can apply for reimbursement about 75% of their organic certification costs through the national organic certification cost-share program. USDA's Agricultural Marketing service has online a lot of resources that can be beneficial for this purpose

Source Anonymous (2016b)

and environment (Damalas and Eleftherohorinos 2011). Several studies have also shown residues of pesticide even on vegetable crops cultivated on organic land. Although pesticides on city surfaces and in city water do not necessarily come from horticultural production in urban areas, when such water and soils are used, it results in high risk. Eiumnoh and Parkpian (1998) from the result of a Bangkok study reported that organo-chlorine and phosphate substances were found to remain in water utilized for irrigation purposes.

1.4 Urban Agricultural Structures in Modern Times

Among metropolitan communities, crops with short cycles and shelf life are more popular while in peri-urban communities, orchards and long cycle crops are the preference (Francesco et al. 2013). Leafy vegetables are a great treasure for human beings as a source of nutrients, vitamins, fibers, phytochemical compounds, etc., in the human diet and they are short cycle crops commonly cultivated in urban communities. Leafy vegetables could be consumed in different methods (fresh or cooked). Leafy vegetables not only play a vital role in human diet but also play a greater role in improve human health and fight the different risky diseases. The increase of public awareness toward the leafy vegetables against junk food especially in urban areas drives the force to consume more leafy vegetables as essential part of a balanced diet (Abul-Soud 2015).

The employment of horticultural practices at home is one of the most common pastimes of individuals in the United States with an industry worth more than $35 billion in 2012 (Anonymous n.d.). Cultivating vegetables in compact spaces (terraces, clear fields, greenhouses or hoop houses) might be an extremely lovely pastime for some aged people who desire to preserve their physical and psychological wellness. For individuals involved in scholarly work, vegetable cultivation is a secure method of rotating intellectual and physical activities, and unwinding (Fig. 1.5). Vegetable growing not only enhances occupational knowledge and skills but also helps to provide training, self-restraint, endurance, and elevation in the standard of living of individuals who perform it repeatedly (Scurtu 2015).

Organic Gardening is gardening without synthetic fertilizer, pesticides, herbicides, and fungicides and many crop possibilities exist such as sweet onions, garlic, peppers, cucumbers, other herbs, and horticultural crops and flowers (pansies, violas, etc.) watermelon, cantaloupe, squash, eggplant, and many more. In recent years, the demand for organic food is gaining popularity (Yue et al. 2011) with a yearly increase in organic food sales worldwide of over $5 billion (Willer et al. 2008). Some reasons why people buy organic foods is because (1) they believe that it is fresh, tasty, environmentally sustainable, and of great benefit to health and of high quality; and (2) they want to stay away from foods whose genetic material has been altered (Demeritt 2002; Schifferstein and Oude Ophuis 1998). Organic farming can help ecosystem and farmland development by building healthy, fertile soils and providing clean water.

Modern farming systems that utilize nutrient-rich water rather than soil for plant nourishment such as hydroponics, aeroponics, and aquaponics (Bridgewood 2003) do not require fertile land in order to be effective. These new modern agriculture systems require less water and space compared with the conventional agricultural systems, employ little or zero use of agricultural chemicals that are not safe for humans consumption or for growers during application (AlShrouf 2017), and can be designed to support continuous production throughout the year (Brechner et al. n.d.). One more advantage of those technologies is the ability to practice the vertical farming production which increases the yield of the area unit (Marginson 2010).

Fig. 1.5 Community garden, (top left); Rooftop garden (top right): Courtesy-Ben Flanner; Compost tea (below left), Vermi compost at the urban farms (below right)

1.4.1 Vertical Farming

Gilbert Ellis Bailey, an expert of geology from America first, used the word "vertical farming" in 1915 (Fabian and Malte 2011). Vertical farming, started and patented by Patrick Blanc, is a geometric method of scaling up production, which decreases area requirements. It involves generally growing plants and rearing animals for food or other food products by artificially aligning them in a vertical position, one above the other (Anirudh and Rekha 2014) in a basic design where soil is not needed (Malisia and Allison 2013). Traditional farming requires soil and uses the area of an acre to grow crops, whereas vertical gardens use a volume of the acre, rising perpendicular to the area, providing additional growth opportunity in 3D area. Vertical systems could be a pile of containers arranged in stacks or several levels of interconnected systems of growing containers populating several acres. These systems work well indoors and outdoors. The structures are reusable, the media is reusable and also compostable. Alternative forms of energy can supply the necessary energy to the pump(s) and the nutrient bath can be developed using organic ingredients.

Vertical layering farming methods have been employed by South American natives for many years, and the flat areas of land on hillsides were used to grow rice in the eastern part of Asia established on the same basic idea (Fabian and Malte 2011). As more and more people are becoming interested in vertical farming, the technologies, designs, and efficiency of such systems are rapidly being upgraded (Schans et al. 2014). Despommier and a group of his trainees in 1999 developed a high-rise concept from rooftop farming after determining that one vertical farm, 30 stories high, could provide food for 50,000 persons. They also theorized that 160 of such systems would be able to supply the entire state of New York with food all throughout the year, without the trepidation of not being able to manage problems associated with harsh weather conditions or drought (Fabian and Malte 2011). In order to solve the challenges associated with the growing world population, food security, and the importance of building soil health, it is anticipated that come 2027, vertical farming would be a dominant trend and it would be financially successful (Zapa 2014).

Vertical farming includes three types of farming (Anirudh and Rekha 2014): 1. The first type is vertical farming done under the surface of the ground, currently practiced in the Netherlands; 2. Vertical farming could be done on very tall story buildings under controlled climatic environment or in open air spaces; 3. The third group entails commercial cultivation in skyscrapers within enclosed structures. All types of vertical farming methods are being tested in a variety of places in different countries of the world (Anirudh and Rekha 2014) such as Singapore, Canada, London, and the U.S.

To run the vertical farm, a framework of support that consists of storage vessels, a water supply system, and other necessary items for plant fertilization are needed. The structure of the soilless vertical system is based on a tower system of a multiple number of pots on a pole with varying levels containing soilless media. The type and number of containers utilized vary with the form and quality of the long, rounded piece of wood, or metal that is used as a form of support (Bielinski et al. 2010). The pull of gravity moves water from the pinnacle of each tower through each strata enabling sufficient water to drain down to the bottom levels (Bielinski et al. 2010). Fertilizer can be applied by slow or fast release fertilizers or through fertigation. The top pots in this design should not be root intensive claiming the majority of nutrients and depriving the lower pots of their fair share and as such should contain plants with lower tolerance to low-light conditions. Heavier fruits or vegetables with moderate to high salinity tolerance are recommended in the ground pot since it serves as a repository for excessive moisture, salt, and fertilizers. Leafy green vegetables and herbs have a short growing season of about 35 days from seed sowing to harvest and so are very well suitable for cultivation on a vertical farm (Boston 2014; McCollow 2014). In addition to herbs and leafy vegetables, eggplant, peppers, and strawberries are great crops for planting in the top 2–3 levels of the vertical tower gardens while for the lower levels, onions, leek, and other crops are satisfactory (Bielinski et al. 2010).

Vertical gardening is practiced for many reasons including for both its ecological and monetary benefits (Chin et al. 2005): for plant enthusiasts and specialists, growing vertically in constrained spaces is an interesting practice; for instructors, vertical garden cultivation presents a creative and exhilarating practical opportunity to teach aspects of science, arithmetic, and finance; for consumers cognizant of healthy eating, vertical planting, built on an organic framework, offers cleaner, fresher vegetables, and fruits with little or no chemical usage; for business-conscious individuals without farmland, vertical gardening provides an alternative within the provision of a market specifically for specialty herbs, vegetables, and fruit production.

This controlled environment provides several benefits, for example, diminished likelihood of disease-pest occurrence and environmental disturbance as well as possibility of maintaining the produce quality and decreasing the amount of water utilized (Schans et al. 2014). Vertical farming system in addition to saving water used in production makes sales and transportation of food products within urban cities easier and less costly. Growing crops in vertical gardens saves land for conventional crop production and is the solution required for conventional crop production. There were more than 12 vertical farms either running or under development in the U.S. as of March 2015 (Birkby 2016).

1.4.1.1 Vertical Farming Production Systems

Vertical farming largely employs the use of soilless culture (commonly referred to as hydroponics). Soilless culture, the most rigorous practice of crop cultivation in the agricultural production systems, is centered on the application of scientific knowledge that is environmentally friendly, and is able to culminate into higher yields equally even in surrounding environment without favorable growing conditions (Gruda 2009). According to Gruda (2009), studies have shown that in contrast to land cultivation, soilless culture allows plant growers produce vegetables, without a deficit in standard. The use of soilless culture techniques in producing vegetables under urban agricultural has led to the avoidance of problems of urban soil contamination, shortage of soil, water, and natural resources besides maximizing the production (Abul-Soud 2015). As a rule, cultivators require substrates that are dependable, reproducible, accessible, simple to mix and manage, and also have chemical and physical characteristics that are suitable for the plant they are cultivating (Klock-Moore 2000).

Some organic substrate constituents that are extensively used are compost, perlite, sand, coconut fiber, pine bark, rice hulls, peat moss, and sphagnum peat. Soilless culture practice must incorporate the four mainstays of sustainable farming (Urrestarazu 2011): (1) Care for the well-being of buyers; (2) Support the farmers' wellness; (3) Protect the environment; and (4) Be advantageous to the agricultural welfare of the community. Soilless culture systems can be grouped into various categories. Two basic groups include the open or closed structure growing systems (Urrestarazu 2011). This categorization best depicts the effectiveness of better nutrient and water management, particularly in the management of environmental emissions and fertigation setups (Urrestarazu and García 2000). Two types of vertical garden systems (open field and indoor) are shown in Fig. 1.6.

Fig. 1.6 Organic vertical gardening in the open field (top) and in door (below)

1.4.2 Hydroponics

Hydroponics is a vital alternative cultivation practice that can be explored by vegetable growers as it can be set up within a building on a scale as little as 400 ft^2, or with the magnitude of a small shed (Anonymous 2016d). Hydroponics is growing methods without soil where roots are either suspended/floating in water or supported with grow media (which solves many soil-born issues), such as rocks or clay or pebbles (Von Seggern et al. 2015) with mineral nutrients added and a lighting structural supply—most often LED lights. Water can be supplied to the media beds from time to time via drip irrigation or other types of irrigation. Excess water and nutrients are then permitted to run through, while extra moisture is included as plants soak them up. The most significant benefit of soilless culture is water conservation as it uses less water than used in traditional agriculture (Chalmers 2004). In a hydroponics system, water levels are constantly monitored to ascertain exact chemical constituents needed are utilized and very little moisture is lost due to evaporation (Birkby 2016). Furthermore, the nutrient solution can also be reused (Anonymous 2011).

Hydroponics is also typically performed in a controlled indoor environment which allows for year-round access to organic, pesticide-free, and fresh daily produce which might not be readily accessible in urban areas especially during winter months (Anonymous 2016c). Ecological factors such as relative humidity and temperature can be calibrated to proportions that will lessen the dangers of microscopic organisms and infection, and would allow easier management of destructive insects or other organisms (Anonymous 2016d). Such systems are very useful and can range from simple setups to highly sophisticate ones. In the hydroponic system, no nutrition pollution is released into the environment; higher nutrient and water use efficiency is attained due to control over nutrient levels (AlShrouf 2017). The hydroponic structure is extremely versatile in that about any plant will develop in hydroponics and deliver steady and significant returns (Fig. 1.7). Knaus (2016) evaluated that for a little baby salad mix business operation (2000–3000 ft^2), 20–30 lbs/sf of produce is likely to be harvested. By and large, reports have demonstrated that the decision of what crops to grow usually depends on the planned production size. Some plants commonly grown in substantial scale hydroponic systems include herbs like basil, cilantro and parsley, and leafy green vegetables such as kale, lettuce, swiss chard, and spinach (McCollow 2014). Other plants that grow well using this method are short duration moisture-loving crops like peas, beans, chives, watercress, scallions, and basil (Prizzia et al. n.d.).

Since hydroponic plants are characteristically grown without pesticide application, it can be cultivated consistent with the requirements necessary for it to be accredited as certified organic nevertheless, there is a layer of intricacy to this procedure because using certified organic manures, fertilizers, and growing media can lead to the development of unwanted biomass can make waste management an issue. A national committee has been created to assess the restrictions surrounding the organic certification of hydroponic operations, and if there is an ease on the present constraints, it will be less demanding for a hydroponic farm to become organic certi-

Fig. 1.7 Lettuce production in hydroponic system (left) Roots development in hydroponic tanks (right): Courtesy-Jim Rakocy

fied, possibly resulting in an increased market for hydroponic products (Anonymous 2016d). Although crops grown in a hydroponic system may appear immaculate to a consumer, some other do not trust that can be as nourishing as crops cultivated in soil (Whoriskey 2015) since they are usually grown in the absence of soil and direct sunlight. Mini hydroponic business operations may require capital of up to $100,000 (Knaus 2016). The procedure is also a laborious and thus requiring a sufficient number of laborers to gather produce especially in the case of leafy green vegetables with a short growing season, and this could become a huge money related impediment for vertical growers producing large scale (Rose 2015). In November 2017, the USDA national organic standards board ruled that hydroponic and aquaponic growers could remain organic certified. It is believed that this decision could increase the number of organic producers as farmers can now grow their produce much easier and faster with good yield without waiting for three years to transition. Nevertheless, the argument persists by many organic growers, that this method is anti-thesis to 'organic' as it does not require the use of soil. In addition, some growers are concerned that plants produced in soilless systems may be less nutritious, weaker and prone to insect pest damage, there by requiring more pesticide application than plants grown in soil. However, as of date, little or no studies have been done to confirm that plants grown in soil are of better or poorer quality those grown in soilless systems and vice versa.

1.4.3 Aeroponics

"In the 1990s, National Aeronautical and Space Administration was keen on finding effective approaches to developing plants in space and instituted the expression "aeroponics", characterized as developing plants in air/mist conditions with next to no water and without soil (Birkby 2016)." Aeroponics is the most productive method of vertical farm cultivation, utilizing about 90% less water than the most effective hydroponic framework (Birkby 2016). The aeroponic system can be used to cultivate vegetables, flowers, and fruits. Plants produced using aeroponics are possibly of a higher nutrient quality and more advantageous as they have been shown to take up more vitamins and nutrients (Birkby 2016). Aeroponic systems are less demanding when it comes to automation even though it makes use of advanced pumps and control structures. Aeroponic farms are also more attractive than their hydroponic counterparts because, in addition to their water conservation properties, they are not prone to disease problems (Boston 2014).

Aeroponics is highly beneficial for growing a lot of plant materials in a limited area, since plants develop vertically, arranged in either a tube-like framework or hanging down in a container, thus reducing the amount of ground space required (Anonymous 2013). The plant root zone area is attached to the middle of the tube (or the base of containers) where water streams past every now and then, apportioning measured quantities of water containing nutrients straight onto the roots (Fig. 1.8). Surplus water that runs out through the base is gathered, and then recycled round the system. This technique makes the best use of the bulk of oxygen reaching the root zone of the plants, helping them assimilate more vitamins and minerals (Anonymous 2013). The vertical aeroponic system in urban areas generally requires sunshine or LED lights, a flat surface of land accessible by road not shadowed by walled structures or mountains, water of suitable quality for agricultural use, and a little measure of electricity (e. g solar power) (Reinhold 2005). Because the aeroponic system is a continuous-cycle in a semi-enclosed to enclosed space, it reduces the agricultural labor into a series of mechanical routine operational tasks which are carried out daily and throughout the year and it needs to be maintained more often (AlShrouf 2017; Reinhold 2005). Organic aeroponic farms would deliver season round higher quality, tastier produce comparable to similar crops grown traditionally, with the smallest quantity of fertilizer. Other benefits include annual crop yields of at least 30 times that of normal agriculture, significant business gains, fixed service hours, not as much back disturbing duties, and properly trained laborers with skills gathered within a short period of time (Reinhold 2005). The leading aeroponics vertical farming company in the United States, Aero Farms, is currently building the largest vertical farm in the nation in New Jersey (Birkby 2016).

1.4.4 Aquaponics

Aquaponics is another alternative for urban agriculturists. Aquaponics merges hydroponics and aquaculture practices in crop production (Perkins 2013). Apart from producing vigorous plants, an aquaponic framework would also rear fish that can be sold through community supply networks (Anonymous 2016a, b, c, d). The fundamental procedure entails adding microorganisms that fix nitrogen to bowls that contain water and fish, and then changes ammonia in fish waste into nitrates (Perkins 2013). As waste build up in water in the aquaculture practice, fish are exposed to a higher level of poisonous substances (Anonymous 2011). Water is then directed to a hydroponic framework where the byproducts from the aquaculture are strained through the plants as essential nutrient solution for growth and the purified water is recycled back and reused by the fish (Fig. 1.9). Aquaponics utilizes just 2% of the water required in conventional farming, in addition to being a closed cycle system with reusable byproducts and as such free from waste (McCollow 2014). New Jersey-based aero-farms reportedly utilize water 95% less water than needed to produce a similar quantity of leafy green vegetables under field conditions (Peters 2015). Aquaponics also does not need as much cleaning amidst harvests, since the microorganisms incorporated often displace the toxic microbes (Boston 2014).

The greatest problem faced by aquaponic business operators is that of filtration: if there is an accumulation of debris, the fish would not survive and the water's chemical distribution will lose the capacity to aid plant development (McCollow 2014). While the cost of rent can become a noteworthy issue for bigger sized operations, a particular aquaponics organization observed that their vegetables developed so quickly that

Fig. 1.8 Aeroponic tower gardens with nutrient tank at the bottom (left) or moving where roots get nutrient spray (right)

Fig. 1.9 Aquaponics system where plants and fish grow together

production balanced out the electricity bill expenses of the high-lease office space (McCollow 2014).

A few persons actively engaged in hydroponic practices have conveyed their doubt. One individual noticed the challenge in ensuring that the needs of the two different systems (the leafy green vegetables and the fish) are met without close observation, to avoid trading-off the nutritional quality of both (Knaus 2016). Consumers are weary of food grown without soil which is fertilized by fish effluent mainly because the idea is foreign to them. One of the most common crops in aquaponics is microgreens, which require hand cutting and packaging which can be incredibly time consuming, if available hands are limited and the yields are large (Von Seggern et al. 2015). Critics also question why the majority of the materials used in an aquaponics system are not sustainable as most systems use components which are made of plastic, metal, PVC, pipe, and synthetic glue (Von Seggern et al. 2015). Another main criticism of aquaponics is the food used to feed the fish as in commercial aquaponics, the food fed to the fish livestock is typically fish food pellets made from smaller bait fish which were grown specifically to be made into fish food as opposed to flakes (Von Seggern et al. 2015). Aquaponic systems can be more complicated than small-scale agriculture where correct setup, monitoring, maintenance, and pH balance are all necessary for growth and day-to-day labor costs for repair and maintenance can come at a significant financial cost (Von Seggern et al. 2015).

1.5 TSU Organic Vertical System Studies

Two research trials were conducted, in fall 2015 and spring 2016, at Tennessee State University (TSU) College of Agriculture, Organic Agriculture Research Farm in Nashville, TN in a hoop house using a vertical garden unit hydroponic system. Ten stacks were set up, each having 22 plants for a total of 220 plants in an area of 75 ft^2. The unit comprised of a ground pot for four plants, four square Styrofoam pots with each holding four plants at the corner, and a small nutrient pot with two plants. Automated system used a 44-gallon tank to supply organic nutrients to plants. Coco fiber and perlite were used as media in pots. Lettuce, chard, kale, mustard green, and coriander were planted in corner of each stacked pot, ground pot, and side of the nutrient pot (Fig. 1.10) (Table 1.2).

Several studies similar to our study have been done. A Dutch agricultural organization, called The Plant Lab, located on a three-story building underground, have grown everything from roses to shrubs to almost every type of plant conceivable such as beans, corn, strawberries, and cucumbers using the vertical growing system (Fabian and Malte 2011). Broward County Driftwood middle school students using a combination of hydroponics and vertical gardening have grown leafy green vegetables, cucumbers, strawberries, and so on (Prizzia et al. n.d.).

Plant Lab's Gertjan Meeuws reported yields thrice the magnitude of a standard greenhouse using 90% less water than a conventional farm and no sunlight in their underground vertical system (Fabian and Malte 2011). In contrast to this, however,

Table 1.2 Leafy greens and herbs grown in vertical garden research trial

Crops	Variety name	Yield (lb/m^2)
Mustard greens	Giant Red	9.3
Kale	Westlander, Lacinto, Meadowlark	3.0
Lettuce	Romaine Lettuce	20.6
Swiss chard	Fordhook Giant	6.6
Coriander (Cilantro)	Sabor	2.0

an experiment by Utami et al. (2011) in Indonesia reported that vertical farming was most appropriate for crop plants that must have intense sunlight like some vegetables and fruits.

Banerjee and Adenaeuer (2014) reported that a study by DLR Bremen estimated the investment, production costs, and market potential of an aquaponic farm, 37 floors high in Berlin to require 3.5 GWh worth of electricity and 80 million liters of water totaling about € 200 million annually. The price of food ranged from € 3.50 to € 4.00 for every kilogram produced and a possible market for about 50–3000 farms was estimated (Banerjee and Adenaeuer 2014). Fabian and Malte (2011), in a Netherland study, reported that consumers are not only requesting more for pesticide-free organic produce but are willing to pay more.

1.5.1 Organic Certification

There is an ongoing debate on the connection between organic cultivation practices and organic vertical gardens in urban horticulture. A large number of agriculturists believe that certified organic farming includes the whole soil and ecological system

Fig. 1.10 Various models of tower gardens for leafy greens, herbs and edible flower production

environment and not just being herbicide or pesticide free (Birkby 2016). Therefore, since soil is absent in vertical gardening cultivation, vertical farms would experience challenges meeting the concept of organic certification, as soil organic activity is absent (Birkby 2016). Nevertheless, the National Organic Program currently permits some organic certification offices to grant accreditation to hydroponic operations if they can demonstrate that they utilize approved organic materials only in their operation and they fulfill other certification standards (Birkby 2016). Though a few national accreditation offices (for example, California Certified Organic Farmers) have granted certification to some hydroponic operations, other organic certification agencies in the region are refusing to affirm hydroponic and other production systems that are not soil based.

1.5.2 Economic Analysis of Vertical Farming

According to Despommier (2012), vertical farms would initially cost a lot and subsequently, and the price would reduce as more people start to request for it as they begin see how it benefits man and the environment (Despommier 2012). Urban green space architects are advancing the vertical farming technology with automatic system controls to minimize labor costs (Rodriguez-Delfin and Martinez 2009). Multinational organizations such as Philips are effectively designing LED lights that are appropriate and affordable for vertical gardens in urban communities (Bhanoo 2016). As this segment continues to expand (Anonymous 2016d), vertical farm growers have observed that plant development increases by 20% as a consequence of utilizing these particular LED lights (Doucleff 2013). In Charleston, U.S., the Environmental Protection Agency has given support to an experiment set out to investigate the plausibility of vertical farming system on a particular site on the city peninsula (Anonymous 2011). For us to really benefit from vertical farming operations, a lot of research still needs to be done on the advantages of the technology to the economy, the environment, and the society.

1.5.3 Benefits of Vertical Farming

Vertical farming is a cost-proficient food production method (Jansen 2011). It can be utilized where there is next to zero land for cultivation and thus it is perfect for kitchen gardening on terraces (Jansen 2011). Vertical farming will encourage organic farming on a substantial level (Anirudh and Rekha 2014). By employing indoor vertical farming methods, grower does not need worry about land-related issues such as pollution, environmental remediation, and zoning (Ryan et al. 2012). Also, the utilization of synthetic pesticides would be reduced using this technology (Anirudh and Rekha 2014). Despommier (2012) noted other advantages of vertical farming to include: produce delivery in all seasons; absence of yield loss due to climatic conditions; utilization of fossil fuels not required, creates work opportunities in urban areas; allows for the water conservation and reuse; and food deterioration due to unnecessary handling is limited. The vertical farming is not affected by climatic variations as it is largely operated under controlled environmental conditions (Germer et al. 2011). Such vertical gardens that are set up inside buildings are protected from extremely hot summer temperatures that may ensue due to climatic variations (Bass and Baskaran 2003). Crops cultivated in vertical gardens are generally simpler to grow and harvest and they are also less likely to be affected by diseases and pests (Utami et al. 2011). Due to the use of soilless media, weeds for the most part do not grow thereby minimizing the costs of labor in that regard. Through the vertical gardening practice, the plant gets sufficient aeration, the aesthetics of the garden are improved. Higher yields are also observed when this growing technique is compared with the traditional cultivation methods. For instance, Chinese cabbage yield produced while utilizing the vertical farming technique was found to be 45 t ha^{-1}, while the median yield of the same crop grown under field conditions was 10–25 t ha^{-1} (Utami et al. 2011).

1.5.4 Challenges in Vertical Farming

Organic vertical farming is however not without challenges. Some challenges include:

Standardization. Lack of standardization and lack of data present a significant obstacle to the establishment of a performance baseline as the majority of the vertical farms interviewed in the study by Jasim et al. (2016) did not keep track of metrics like monthly energy usage, water usage, precise yield, waste diversion, air quality, and amount of pesticide or fertilizer use. In addition to many farms lacking quantitative data, the ones that did have information reported in a non-standardized way. An example is water usage—one farm reported watering for 1 h/day, while another reported that 8.5 gallons were used for a head of lettuce (Jasim et al. 2016).

Produce quality. Numerous studies confirm that soilless culture systems (SCSs) enable farmers to grow their vegetables without reduction in quality when compared to soil cultivation. Be that as it may, utilizing soilless culture does not just bring about the harvest of vegetables of high standard without proper cultural management peculiar to that particular system (Gruda 2009).

Light and heating. Light is a major challenge, because light-emitting diodes have to take the place of natural sunshine and artificial heating may result in spiked energy costs, operating costs, capital and general investment costs especially in large-scale commercial production thereby impacting profitability (Fabian and Malte 2011). They reported (Fabian and Malte 2011) that to produce certain crops such as wheat on vertical farms in the U.S., eight times more electricity would be required by the energy plants to power lighting. When a renewable power supply is required, it becomes more of a challenge to do so.

Crop Species and pollination needs. Vertical farming cultivation favors the production of vegetables or crops that develop rapidly such as herbs and salad items. Leafy green vegetables that grow at a slower rate and grains do not yield as much profit. As insects are unavailable on a vertical farm, plant pollination is usually done by hand, and in such situations, more manpower and effort is a requirement (Birkby 2016).

Other barriers. Plant development could be hindered when nutrient supplementation, as well as light introduction, is not closely checked and as such the problems related to finding an area that gives the essential requirements for growing indoors, water, air, ventilation, and other basic utilities may arise (Ryan et al. 2012). Furthermore, the cost of the land or building in an urban site may be expensive if it is not in an abandoned or derelict area. Insect pests such as aphids are challenge in organic vertical farming system. In such situations, the plants may be periodically sprayed as needed with Organic Materials Review Institute (OMRI) approved organic pesticide options such as DiPel, Neem Oil, M pede, PyGanic, Monterey BT, Safer Insect Killing Soap, Mycotrol O and Suffoil X to discourage pests (Nwosisi et al. 2017).

1.6 Recommendations

Analysis is necessary (Bhanoo 2016) on the environmental, economic costs, social inclusion, and consumer acceptance of vertical farms and farm produce. More research is needed to determine the environmental effect of hydroponics in comparison to traditional leafy green cultivation on the field. The degree to which vertical gardens can reduce the heat waves in urban communities today cannot be known for sure without additional experiments comparing the effect of climate change in urban centers without and with the implementation of this technology (Bass and Baskaran 2003). Also, further studies are required to explore the replacement effect of chemical nutrient solutions by organic nutrient solution (Abul-Soud 2015). Banerjee and Adenaeuer (2014) in their study on the economics of vertical farming found that there is a possibility that vertical farm produce may become so much more costly for it to become practically applicable (Banerjee and Adenaeuer 2014). In order to bring that down, more studies need to be done on a small and larger scale, to find more effective methods of production as well as the modification and merge of vertical farm network together with other ventures are essential (Banerjee and Adenaeuer 2014). Further research is needed to account for the significant number of urban and peri-urban growers practicing vertical farming today, and the various roles of food production in and around urban communities and the impact of other issues like waste management, flood water, greening, relaxation, instruction/training, community development, etc., on vertical farming systems.

1.7 Conclusion

Urban horticulture and vertical farming are major food and income source for many urban settlers around the world, and when merged with other technologies, it can assist urban territories in absorbing increasing populations while providing food security. Vertical farming is sustainable. The financial commitment can be small or large depending on the needs of the individual or farm. Due to an expansion in food insecurity, and a reduction in accessible funds and land spaces, additional research in large- and small-scale vertical farming has become essential to tackle food insecurity and to allow a means for low-earning families to obtain their fundamental nutritional needs. With the vertical farming technique, individuals can produce vegetables and fruits for themselves on a little land area with limited expertise.

1.8 Further Studies

Further studies would help understand why urban horticulture is essential as a source of income and how this varies with wealth, gender, and residential areas. Government policies relevant to priorities of urban horticulture growth and development should be revised and established, e.g., in relation to crop science/technology, market structures, and food security with end consumers.

References

Abul-Soud, MA, Emam, MSA, Abd El-Rahman Noha, G (2015) The potential use of vermi- compost in soilless culture for producing strawberry. Int J Plant Soil Sci 8(5):1–5

Agnes AD (2014) African urbanization trends and implications for urban agriculture. In: Ulf M, Kristin FB (eds) Urban and peri-urban agriculture for food security in low-income countries—challenges and knowledge gaps. SLU-Global Report publication, Lund, Sweden, pp 6–8. Available from: https://www.slu.se/globalassets/ew/org/andra-enh/uadm/global/rapporter/slu-global-report-2014-4-urban-and-peri-urban-agriculture-for-food-security-webb.pdf

Albert AA (2012) Urban horticulture and community economic development of lagging regions. In: Luna Maldonado AI (ed) Horticulture. InTech, ISBN: 978-953-51-0252-6. Available from: https://cdn.intechopen.com/pdfs-wm/31554.pdf

AlShrouf A (2017) Hydroponics, aeroponic and aquaponic as compared with conventional farming. Am Sci Res J Eng Technol Sci 27(1):247–255. Available from https://www.asrjetsjournal.org/index.php/American_Scientific_Journal/article/viewFile/2543/1028

Anirudh G, Rekha B (2014) Recent trends in agriculture: vertical farming and organic farming. Adv Plants Agric Res 1(4):00023. Available from https://www.researchgate.net/file.PostFileLoader.html?id=55cc81445dbbbdb5ce8b45ec&assetKey=AS%3A273837743116289%401442299413989

Anonymous (2011) Chaleston vertical farm design feasibility study. Clemson University Institute of Applied technology, Clemson SC, p 210. Available from http://media.clemson.edu/public/restoration/iae/vfarm/cuiae_epa_vfreport30mb.pdf

Anonymous (2013) From out of the blue, green farming. Recirculating farms coalition, p 32. Available from http://www.recirculatingfarms.org/wp-content/uploads/2013/06/RFCreport_FINAL-FINAL.pdf

Anonymous (2014) World urbanization prospects: the 2014 revision, highlights. United Nations, Department of Economic and Social Affairs, Population Division. (ST/ESA/SER.A/352). Available from https://esa.un.org/unpd/wup/publications/files/wup2014-highlights.Pdf

Anonymous (2016a) Fruit and tree nuts lead the growth of horticultural production value. United States Department of Agriculture, Economic Research Service. Available from https://www.ers.usda.gov/data-products/chart-gallery/gallery/chart-detail/?chartId=76958

Anonymous (2016b) USDA urban agriculture toolkit: summary of key resources, p 4. Available from https://www.usda.gov/sites/default/files/documents/urban-agriculture-toolkit.pdf

Anonymous (2016c) Aquaponics FAQ. Endless food systems, Forestburg, TX. Available from https://www.endlessfoodsystems.com/faq/

Anonymous (2016d) Differentiated cost of production in the Northwest: an analysis of six food categories. Leafy greens. ECOTRUST, Portland Oregon, p 52. Available from https://ecotrust.org/media/CFFP_leafy_greens_8_16_16.pdf

Anonymous (n.d.) USDA definition of specialty crops. United States Department of Agriculture, Agriculture Marketing Service, p 11. Available from: https://www.ams.usda.gov/sites/default/files/media/USDASpecialtyCropDefinition.pdf

Banerjee C, Adenaeuer L (2014) Up, up and away! The economics of vertical farming. J Agric Stud 2(1):40–60

Bass B, Baskaran B (2003) Evaluating rooftop and vertical gardens as an adaptation strategy for urban areas. NRCC-46737. National Research Council Canada. CCAF Report B1046, p 111. https://www.nps.gov/tps/sustainability/greendocs/bass.pdf

Bhanoo SN (2016) Vertical farms will be big, but for whom? Fast Company. Available from http://www.fastcompany.com/3039087/elasticity/vertical-farms-will-be-bigbut-for-who. Accessed 4 Feb 2016

Bielinski MS, Teresa PS-D, Shawn CA (2010) Solutions for small farmers and home gardens: building a low-cost vertical soilless system for production of small vegetable and fruit crops. Horticultural Science Department UF/IFAS Extension. HS1186. Available from http://edis.ifas.ufl.edu/pdffiles/HS/HS118600.pdf

Birkby J (2016) Vertical farming. ATTRA Sustainable Agriculture. NCAT IP516, p 12

Boston HM (2014) Indoor farms: making light work of city dining. The Economist. Available from http://www.economist.com/blogs/babbage/2014/04/indoor-farms

Brechner M, Both AJ, Staff CEA (n.d.) Hydroponic lettuce handbook. Cornell Controlled Environment Agriculture, pp 504–509

Bridgewood L (2003) Hydroponics: soilless gardening explained. The Crowood Press Limited, Ramsbury, Marlborough, Wiltshire

Chalmers GA (2004) Aquaponics and food safety. Lethbridge, Alberta

Chin L, Yadong Q, Mila B (2005) Vertical gardening. Circular—Sustainable plant and animal production systems. Southern University and A&M College System Agricultural research and extension center, Baton rouge, LA. No 301, p 2. Available from https://www.researchgate.net/publication/270511103_Vertical_Gardening

China Development Research Foundation (2010) Trends in urbanisation and urban policies in OECD Countries: what lessons for China? China development research foundation. Organization for economic co-operation and development (OECD) publishing, p 219. Available from https://www.oecd.org/urban/roundtable/45159707.pdf

Cohen M, Garrett J (2009) The food price crisis and urban food insecurity. IIED, London, UK

Damalas CA, Eleftherohorinos IG (2011) Pesticide exposure, safety issues, and risk assessment indicators. Int J Environ Res Public Health 8(5):1402–1419

Demeritt L (2002) All things organic 2002: a look at the organic consumer. The Hartman Group, Bellevue, WA

Despommier D (2012) Advantages of the vertical farm. In: Rassia ST, Pardalos PM (eds) Architecture: impacts on health, springer optimization and its applications, vol 16. Springer, pp 269–275. Available from https://www.researchgate.net/profile/Dickson_Despommier2/publication/225262607_Advantages_of_the_Vertical_Farm/links/564f0c4d08ae1ef9296e3d4a.pdf

Doucleff M (2013) Vertical pinkhouses: the future of urban farming? NPR: The Salt. Available from http://www.npr.org/sections/thesalt/2013/05/21/185758529/vertical-pinkhouses-the-future-of-urban-farming

Drescher AW, Laquinta D (1999) Urban and peri-urban agriculture: a new challenge for the UN Food and Agriculture Organisation (FAO). FAO, Rome

Eiumnoh A, Parkpian P (1998) Impact of peri-urban vegetable production on soils and water: a case of Bangkok plain, Thailand. Paper presented at the Peri-urban Vegetable production in the Asia-pacific region for the 21st century, 29 Sept–1 Oct, Kasetsart University, Bangkok, Thailand, p 24

Eva De L (2010) Urban agriculture as a strategy for poverty reduction in Uganda: the case of Lira Municipality. A master's thesis, Universiteit Gent, p 133. Available from http://lib.ugent.be/fulltxt/RUG01/001/491/161/RUG01-001491161_2011_0001_AC.pdf

Fabian K, Malte EK (2011) Vertical farming: can urban agriculture feed a hungry world? Spiegel online news International, p 4. Available from http://communitycouncil.ca/sites/default/files/15-Druckversion-VerticalFarming_CanUrbanAgricultureFeedaHungryWorld.PDF

FAO (2008) Urban agriculture for sustainable poverty alleviation and food security. Final Draft. Available from http://www.fao.org/fileadmin/templates/FCIT/PDF/UPA_-WBpaper-Final_Draft-3_October_2008-FG-WOB.pdf

Fassler J, Cox K (2017) It's the end of organic as we know it. New food economy, New York City. Accessed online April 4, 2018 from https://newfoodeconomy.org/hydroponics-farm-organic-nosb-vote/

Francesco O, Remi K, Remi Nono-W, Giorgio G (2013) Urban agriculture in the developing world: a review. Agron Sustain Dev 33:695–720

Germer J, Sauerborn J, Asch F, de Boer J, Schreiber J, Weber G, Muller J (2011) Skyfarming an ecological innovation to enhance global food security. Journal Fur Verbraucherschutz Und Lebensmittelsicherheit 6(2):237–251. http://doi.org/10.1007/s00003-011-0691-6

Gruda N (2009) Do soilless culture systems have an influence on product quality of vegetables? J Appl Bot Food Qual 82:141–147. Available from http://edoc.hu-berlin.de/oa/articles/reMpRZba9vp/PDF/208DDKUFxVNwI.pdf

Jansen T (2011) Farm technology. Protecting food security through adaptation to climate change in Melanesia. Live & Learn Environmental Education, Honiara Solomon Islands. Available from https://www.humanitarianresponse.info/system/files/documents/files/Farm%20Technology.pdf

Jasim A, Tracy B, Michelle B, Ezzeddine M, Halperin G, Harutyunyan A, Levinson A, Macdonald K, Miller K, Peterson S, Sullivan D, Tavoraite D, Vera K, Yuan M, DesRoches S (2016) Sustainability certification for indoor urban and vertical farms: a sustainable approach to addressing growth in vertical farming. Association for vertical farming. M.Sc. students Capstone Workshop. The Earth Institute Columbia University. Available from http://sustainability.ei.columbia.edu/files/2016/01/Sustainability-Certification-for-Indoor-Urban-and-Vertical-Farms.pdf

Katherine HB, Martin B, Alison M-C, Joe N, Jac S, Terri B, Peter M (eds) (2002) Urban agriculture and community food security in the United States: farming from the city center to the Urban Fringe. Urban Agriculture Committee of the Community food security commission. Available from http://ocfoodaccess.org/wp-content/uploads/2013/08/Urban-Agriculture-Food-Security_CFSC-2002.pdf

Kennedy G, Nantel G, Shetty P (2004) Globalization of food systems in developing countries: a synthesis of country case studies. In: Globalization of food systems in developing countries: impact on food security and nutrition. Food and Nutrition Paper, vol 83, pp 1–25. FAO, Rome, Italy

Kerr A (2016) A cornucopia of categories for crops. USDA California Climate Hub. University of California, Davis. Available from http://caclimatehub.ucdavis.edu/2016/07/21/a-cornucopia-of-categories-for-crops/

Klock-Moore KA (2000) Comparison of salvia growth in seaweed compost and bio-solids compost. Compost Sci Utilization 8:24–28

Knaus D (2016) Interview by Noah Enelow. Hydroponic and organic farming for local markets: Market trends and opportunities

Lee-Smith D (2010) Cities feeding people: an up-date on urban agriculture in equatorial Africa. Environ Urbanization 22

Lovell ST (2010) Multifunctional urban agriculture for sustainable land use planning in the United States. Sustainability 2:2499–252. https://doi.org/10.3390/su2082499

Malisia W, Allison K (2013) Onward & upward: filling the hunger gaps with vertical farming. Hort 408 Literature Review. Clemson University, p 50. Available from http://www.clemson.edu/cafls/vincent/articles/lit_review.pdf

Marginson S (2010) Aero-farms urban agriculture system: Less space, less water, and no pesticides. Available from http://www.gizmag.com/aerofarms-urban-agriculture/15371/

McCollow K (2014) Aquaponics revives an ancient farming technique to feed the World. Newsweek. Available from http://www.newsweek.com/2014/05/23/aquaponics-revives-ancient-farming-technique-feed-world-251020.html

Moustier P (1999) Definations and outlines of peri-urban agriculture in sub-Saharan Africa. CIRAD, Colloques, pp 17–29

Nugent R (2000) The impact of urban agriculture on the household and local economies, pp 67–97. Available from http://wentfishing.net/farmlit/Theme3.pdf

Nwosisi S, Nandwani D, Saidullah C (2017) Organic vertical gardening of vegetable production for urban and semi-urban growers. Acta Hortic (1189). ISHS 2017. In: M. Papafotiou et al (eds) Proc. VI Int. Conf. on Landscape and Urban Horticulture. https://doi.org/10.17660/ActaHortic.2017.1189.76

Perkins C (2013) A Koi-Fueled nursery in New Orleans yields tasty profits. Wired.com: Conde Nast Digital. Available from http://www.wired.com/2013/12/veggi/

Peters A (2015) Farmed here wants to bring a vertical farm to your city. Fast Company: Co. Exist. Available from http://www.fastcoexist.com/3053217/farmedherewants-to-bring-a-vertical-farm-to-your-city

Philippe T, Hubert de B (2006) Urban horticulture. In: René, van. V. (ed) Cities Farming for the future; urban agriculture for green and productive cities, vol 11. RUAF Foundation, Netherlands, IDRC, Canada and IIRR publishers, Philippines, p 460. Available from http://citeseerx.ist.psu.edu/viewdoc/download?doi=10.1.1.732.7826&rep=rep1&type=pdf

Popkin BM (1999) Urbanization, lifestyle changes and the nutrition transition. World Dev 27(11):1905–1916

Prizzia A, DelValle T, Wilber W, McCoy S, DeValerio J, Skvarch E, Spriggs S, Simonne A, Thralls E, Hunsberger A, Orwat M, Christensen C, Ozores-Hampton M, Popenoe J, Brew K (n.d.) Grow to learn. In: Treadwell D, Brew K, Sydney PB, Prizzia A (eds) UF/IFAS University of Florida department of agriculture and consumer sciences, pp 26–28. Available from http://uffnp.org/wordpress1/wp-content/uploads/2015/09/FDACSFNW_Grow-to-Learn_Book-LOW-RES.pdf

Ravallion M, Chen S, Sangraula P (2007) New evidence on the urbanization of global poverty. Popul Dev Rev 4:667. https://doi.org/10.2307/25487618

Redwood M (2009) Agriculture in urban planning: generating livelihoods and food security. Earthscan, London, UK

Reinhold Z (2005) The vertical aeroponic growing system. Synergy International, Inc. Sausalito CA, p 13. Available from http://www.synergyii.com/aeroponic/VAP.pdf

Rodríguez-Delfín A, Martínez PF (2009) Proceedings of the international symposium on soilless culture and hydroponics, Lima, Peru, 25–28 August 2008. Commission Plant substrates and soilless culture. International Society for Horticultural Science, p 410

Rose J (2015) Green pie in the Sky? Vertical farming is on the rise in Newark. The Salt, NPR. Available from http://www.npr.org/sections/thesalt/2015/08/05/429345848/greenpie-in-the-sky-vertical-farming-is-on-the-rise-in-newark. Accessed 10 Feb 2016

Ryan A, Matt H, Vanessa R (2012) Starting a farm in your city: transforming vacant places, vol 1. Green opportunities in growing industries. Delta Institute GO Guides, p 42. Available from http://delta-institute.org/delta/wp-content/uploads/UrbanAgGoGuide_FINAL-for-Web1.pdf

Satterthwaite D, McGranahan G, Tacoli C (2010) Urbanization and its implications for food and farming. Philos Trans R Soc B Biol Sci 365(1554):2809–2820

Sawio CJ (1994) Who are the farmers of Dar-es-Salaam. In: Egziabher AG, Lee-Smith D, Maxwell DG, Memon PA, Mougeot LJA, Sawio CJ (eds) Cities feeding people: an examination of urban agriculture in East Africa. IDRC, Ottawa

Schans Van der JW, Renting H, Rene VV (2014) Innovations in urban agriculture, vol 28. Urban Agriculture Magazine, p 10. Available from http://www.ruaf.org/sites/default/files/p3-12.pdf

Schifferstein HN, Ophuis PAO (1998) Health-related determinants of organic food consumption in the Netherlands. Food Qual Prefer 9(3):119–133

Scurtu I (2015) Vegetable growing—hobby and benefit for aged person health. Curr Trends Nat Sci 4(7):13–22. Available from http://old.upit.ro/uploads/revistastiinte/CTNS%20vol%204%20issue%207/Paper%202.pdf

Smit J, Ratta A, Nasr J (1996) Urban agriculture: food, jobs and sustainable cities. United Nations Development Programme, New York, p 302

Urrestarazu M (2011) State of the art and new trends of soilless culture in Spain and in emerging countries. In: International symposium on growing media, composting and substrate analysis, vol 1013, pp 305–312

Urrestarazu M, García M (2000) Modeling electrical conductivity management in a recirculating nutrient solution under semiarid conditions. J Plant Nutr 3:457–468

Utami SNH, Jayadi R (2011) Vertical gardening for vegetables. In: International symposium on sustainable vegetable production in Southeast Asia, vol 958, pp 195–202

Van Leeuwen E, Nijkamp P, de Noronha Vaz T (2010) The multifunctional use of urban greenspace. Int J Agric Sustain 8(1–2):20–25

Van Veenhuizen R, Danso G (2007) Profitability and sustainability of urban and peri-urban agriculture. FAO agricultural management. Marketing and finance occasional paper no 19. FAO, Rome

Von Seggern L, Jillian S, Andrew Z, Frank R, Roberto Queiroz de Abreu Sodre L (2015) Urban farming-The black pearl gardens. Dow sustainability Fellowship Programmes, University of Michigan. Available from http://sustainability.umich.edu/media/files/dow/Dow-Black-Pearl-Garden.pdf

Whoriskey P (2015) Are these beautiful lettuce and tomatoes really organic? The debate over an agricultural innovation. Washington Post. Available from https://www.washingtonpost.com/news/wonk/wp/2015/04/24/theres-a-raging-debate-over-whetherthese-types-of-lettuce-and-tomato-are-really-organic/

Willer H, Yuseffi-Menzler M, Sorensen N (2008) The world of organic agriculture: statistics and emerging trends 2008. International Federation of Organic Agriculture Movements (IFOAM), Germany

Yue CY, Tong C (2009) Organic or local? investigating consumer preference for fresh produce using a choice experiment with real economic incentives. HortScience 44(2):366–371

Yue C, Dennis JH, Behe BK, Hall CR, Campbell BL, Lopez RG (2011) Investigating consumer preference for organic, local, or sustainable plants. HortScience 46(4):610–615

Zappa M (2014) Emerging agriculture technologies that will change the world. Business Insider. Available from www.businessinsider.com/15-emergingagriculture-technologies-2014-4

Chapter 2
Sustainability of Living Wall Systems Through An Ecosystem Services Lens

F. Larcher, L. Battisti, L. Bianco, R. Giordano, E. Montacchini, V. Serra and S. Tedesco

Abstract The use of plants for shading the buildings having a pleasant ornamental effect is known for a long time. During the last decade, ecosystem services approach opens the way to analyze the multiple advantages that greening, included living walls, can provide to the urban environment. Regarding living walls, the main environmental benefits are heat island effect mitigation, noise pollution insulation, heating and cooling energy demand reduction, absorption of particulate matters, and enhancement of biodiversity. Living walls can also have a role in ameliorating physical and mental health and well-being and can be included in the urban horticulture initiatives both for food production and education purposes. Needing for recycled and low energy products for a greater environmental sustainability is also required in living wall systems (LWS). Therefore, the life cycle approach is considered as useful. The chapter is a critical review aimed at analyzing the environmental, social and economic sustainability of LWS for urban design. The multidisciplinary approach, carried out by agronomists and architects, allows to understand the problems and solutions for the improvement of the use of these systems in our cities. The research group is working on sustainable green wall systems that provide architectural and environmental benefits through the use of independent modular units, renewable materials, environmental friendly substrates, and native plants.

Keywords Acoustic performance · Air quality
Building energy efficiency · Cultural service · Ecosystem service
Green wall · Mitigation of UHI · Urban design

F. Larcher (✉) · L. Battisti
Department of Agriculture, Forest and Food Sciences, University of Turin, Largo Paolo Braccini 2, 10095 Turin, Grugliasco, Italy
e-mail: federica.larcher@unito.it

L. Bianco · V. Serra
TEBE Research Group, Energy Department, Politecnico di Torino, C.so Duca degli Abruzzi 24, Turin, Italy

R. Giordano · E. Montacchini · S. Tedesco
Department of Architecture and Design, Politecnico di Torino, Viale Pier Andrea Mattioli 39, 10125 Turin, Italy

D. Nandwani (ed.), *Urban Horticulture*, Sustainable Development and Biodiversity 18, https://doi.org/10.1007/978-3-319-67017-1_2

2.1 Introduction

Speaking about the sustainability of the urban environment and the role of greening, the multiple use of different technical and ecological solutions must be the future approach for planners and designers. The holistic approach and multidisciplinary teams of research are needed to provide a complete knowledge of the different urban ecosystems, in order to achieve the best long-term result.

The green infrastructures network in the European cities and Metropolitan areas includes a lot of nature-based solutions devoted to improve both the urban natural capital and the quality of life of citizens. Among these solutions, green corridors, parks and gardens, green roofs and green walls can be included.

The aim of the following paragraphs is to make the point about the green walls, starting from the description of the different types and technologies, then moving to the analysis of the ecosystem services provided. In particular, a literature review of papers published during the last 15 years has been done with a particular attention on the European context.

2.2 Living Wall Systems

2.2.1 Classification and Definition

The vertical greening system, also named green wall or vertical garden, can be classified into two main types according to their growing method: green façades and living walls. Green façades use climbing or hanging plants covering one vertical building surface. This solution often includes a support framework, but the plants are rooted in the soil close to the frame or near the building façade, or eventually in grounded pots. Living walls are solutions made up of different materials and technologies that allow the growth of many plant species directly on the vertical surface.

In this context, the present contribution focuses on this second system, the living wall system (LWS). The first and most famous example is the patent "Mur Vegetal" developed in 1988 by the French botanist and designer Patrick Blanc (Fig. 2.1). This system is composed of a double layer of geotextile synthetic felt assembled on waterproof PVC panel. Plants grow in special pockets cut into felt. Water and nutrients are provided through a fertirrigation system.

In last decades, different LWS were developed. The international scientific literature does not provide a univocal classification system. The classification proposed by Weinmaster (2009) identifies two types: the hydroponic system, where irrigation is the delivery of water and nutrients to the plant roots, and the modular box system, where plastic or metal containers are filled with growing medium and 6–20 plants. According to Loh (2008), LWS can be classified into three systems: "panel system"—pre-vegetated panels jointed to a structural frame and equipped

Fig. 2.1 Example of "Mur vegetal" designed by Patrick Blanc, Musée du quai Branly (Paris)

with integrated irrigation system; "felt system"—consisting of a series of felts and a waterproof barrier, attached to a substructure; and "container and/or trellis system"—containers with plants that climb onto trellises. Moreover, Ottelé (2011) proposed a classification of LWS, including only modular system based on plastic planter boxes; foam substrate with steel baskets as support; and LWS based on several felt layers, for supplying water and nutrients to plants. Shaikh et al. (2015) classifies LWS in "modular living wall"—panels with substrate to sustain plants; "vegetated mat wall", referring to the "Mur Vegetal"; and "biofiltration", developed to biofiltering air and providing thermal regulation.

The classification system proposed in this chapter resumes the one described by Manso (2015) that systematizes LWS characteristics and technologies based on international peer-reviewed articles and patents available in the worldwide. The LWS classification system then includes:

- *Continuous systems*, characterized by: plants inserted into felt pockets; hydroponic technique (plants can obtain nutrients through irrigated water).
- *Modular systems*, characterized by: elements with specific dimension; hydroponic technique; presence of some kind of substrate (plants can obtain nutrients through the growing media).

Continuous systems and modular systems support a greater diversity of plant species and require irrigation (Figs. 2.2 and 2.3).

The continuous systems are based on the absorption capacity of water by some materials and on the capacity of plants to develop a superficial root system. Usually, they are made of a synthetic fabric coating, in which some pockets are realized (through some horizontal cuts of the fabric) to contain the plants. The continuous systems are on site assembled, so may not be pre-cultivated in the nursery; these systems have rather high costs of installation and maintenance, but they allow a unique design and pattern over the building envelope choices, thanks to the possibility of use different plant species.

Fig. 2.2 Example of continuous LWS located in Amsterdam in a fitness center

Fig. 2.3 Example of Modular Living Wall System located near Milan (Italy) in a shopping mall

The modular systems are made up of a wide range of materials and elements of various thicknesses, dimension, covering the wall systems. The main features of these systems are

- easy to install;
- have ready-pattern (the modules are pre-vegetated in the nursery);
- easy to replace for maintenance.

The modules are generally made of plastic, geotextile felts, and metal, with differences in their composition, weight, and assembly.

Nevertheless, a knowledge of parameters related to LWS life cycle, according to the mentioned life cycle approach, is a key issue if aimed at choosing the suitable products and systems according to uses, time, climate as well as mandatory requirements included in building code and regulations. Some useful parameters for decision of which LWS is more appropriate to each project are described in Table 2.1.

LWS vary enormously and the design of their elements is constantly evolving, to obtain more efficient solutions and to improve environmental and agronomical performances over the LWS life cycle stages (design, manufacturing, on-site assembling, maintenance).

Table 2.1 List of the main parameters, their description and calculation unit useful for describing LWS

Parameter	Description	Unit
Shape, sizing, thickness	The technological system. The supporting plant framework and sizing affect the plant selection	cm
Weighting	The dry weight and the water saturation weight of an individual module; weight affects the system handling and the framework sizing aimed at carrying the LWS load	kg/m^2
Water consumption	The average seasonal/annual water consumption	l
Plants	The number of plants placed in the LWS per square meter	N/m^2
Principles of growing and conceptions	The use of hydroponic system or substrate to provide water and nutrients to plant roots	Hydroponic/substrate system
Cost	The installation cost; it depends on system conception and material used	€/m^2
Maintenance frequency	The amount of maintenance necessary for pruning, plant species replacement or panel replacement	Time frame

2.2.2 *Plants*

Climbing species are used since a long time for ameliorating the aesthetic aspect of buildings. LWS are not based on a limited range of climbing flora to the same extent as green façades, but allow a greater number of species to be planted on the wall surface (Francis and Lorimer 2011). The continuous system takes advantage of the natural *habitus* of many plant species, with endless possibilities in design and aesthetic results (Weinmaster 2009). In the modular LWS, more limits for plant growing and design effects must be considered.

Looking at the literature, really few papers are devoted to analyze the differences in plant species and their performances in LWS. Patrick Blanc, in his first experience, was inspired by the epiphytes, air plants, that are plants living in the rainforest on other plants and trees. Also, the use of climbing species (naturally moving vertically on some kind of support) is very common. Pérez et al. (2014) in their review collected a list of plant species cited in scientific trials. The most frequent species were *Parthenocissus tricuspidata, Parthenocissus quinquefolia, Lonicera japonica, Clematis* sp., *Hereda helix, Jasminum, Pyracantha, Wisteria sinensis* (as climbers); *Zoysia japonica, Euphorbia lomi, Ophiopogon japonicus, Tradescantia spathacea, Phyllanthus myrtifolius*, and several other shrubs and herbaceous perennials.

More recently, some interesting papers tested the use of Mediterranean shrubs (Devecchi et al. 2013), evergreen perennials (Mårtensson et al. 2014; Bianco et al. 2016) (Fig. 2.4), and edible species (Mårtensson et al. 2016).

Fig. 2.4 The use of evergreen perennial plants for greening a Business Building in London (UK)

Concerning the selection of plants for LWS, some authors suggested to improve the knowledge on “synurbic” species (Francis and Chadwick 2012) and xerophyte species (Benvenuti et al. 2016), considering that these kinds of plants can provide a good ornamental effect together with a water and costs saving.

The use of a suitable plant species, together with the technological frame and the other materials, is a fundamental choice. In particular, Sheweka and Mohamed (2012) suggested to focus the attention on the color and pattern of leaves, density of the foliage, wind penetrability, angle of the foliage exposure, and number of overlapped layers. Also, the canopy persistence during seasons (deciduous or perennial), and the percentage of covering and evenness of the entire wall have to be considered. Moreover, the thickness, composition, and physical conditions of the substrate (porosity and humidity) can have an important role in LWS performances.

2.2.3 Layers, Materials, and Devices

As described in the previous paragraph, there are two general living walls types: continuous felt covered hydroponic panels and modular, which can be hydroponic, or substrate based, or both. The environmental impact depends on the materials used, their durability over time, their potential for recycling, plants replacement frequency, and water consumption. The main LWS' layers, materials, and devices and their sustainability performance in a life cycle perspective are described below.

Frame

Living walls need a frame (horizontal guides, vertical beams or latticework) to be jointed at a certain distance to the envelope. The materials generally used are aluminum, galvanized steel, stainless steel, or some other non-rusting metal. The use of alternative materials, such as recycled plastic (HDPE), hardwood (with FSC certificate), or coated steel, is recommended (Ottelé et al. 2011).

Materials

Continuous LWS are made of waterproof membrane, which is jointed to a frame and added with permeable and flexible synthetic textiles and felts. Recycled fibers such as polyamide, polyethylene, and polypropylene are used, because they give structural support and are long-lasting, even under constant moisture. Plants are placed between felt layers in hand-cut pockets and roots can grow and ramify through the entire system.

Modular LWS use planting containers (e.g., pots, trays, vessels, planter tiles, trellis panel etc.) made of plastic or metal, or synthetic fibers bags. The substrate is packed into the containers or placed into the fiber bags. All these modular systems require an own framework to be fastened to the wall. They are easy to install, replace, and remove.

The durability and the recycling potential of living walls are important parameters too. According to Ottelè et al. (2011), the duration of living wall systems varies between 10 years for the panels based on felt layers, and 50 years for the ones based on planting containers. Moreover, felt layers are difficult to recycle and their environmental disposal impact is similar to the realization one.

Hydroponic versus substrate based systems

The LW hydroponic system can be either large panel or continuous systems, in which plants do not require soil to grow. Panels are generally made up of recycled synthetic fibers and absorbent felts where plants are put. Inert materials (e.g., horticultural foam, perlite, vermiculite, felt mat) can be used for roots holding (Weinmaster 2009). The advantage of the hydroponic system is that there is no structural decay of the growing medium, so the water and nutrients supply is better controlled.

Substrat-based systems typically use modular elements that require the use of a growing medium. The growing medium allows the plant anchoring and facilitates water, air, and nutrient access. At present, the following organic and inorganic growing media are commonly used: coco-coir fiber, peat potting soil, and rock wool. Usually, light materials (e.g., perlite, vermiculite) are mixed in the medium in order to improve its porosity and to reduce the weight of the panel (Manso and Castro-Gomes 2015).

Some authors compare some typologies of growing medium based on their environmental impact in the production phase: coco-coir, rock wool, peat, and potting soil (Weinmaster 2009). According to the study, the best choice is the coco-coir, a waste product from the manufacture of ropes and matting. The peat, which is the decomposed organic matter of semiaquatic plants, is a nonrenewable material. Some researchers confirm the environmental and economic impact of substrates realized in peat and perlite and show the need to develop alternative substrates (Masaguer-Rodríguez et al. 2015; Larcher and Scariot 2009; López-Cuadrado et al. 2006). Recent studies (López-Rodríguez et al. 2016) are focused on new sustainable growing media based on organic residues (composted pine bark, coconut fibers, rice husks, wood fibers).

Irrigation and plant nutrition

Plants on living walls require regular and reliable water supply for their survival. The water needs depend on the species used, the climatic conditions, the exposure, and the type of system. Automated irrigation system is always necessary and avoids plant losses due to the inconsistent moisture management (Bit 2012).

Two types of irrigation systems are available: recirculating and direct irrigation. In a recirculating system, the source of water is an irrigation tank: the water is pumped from the tank to the plants and the excess drainage is fed back to the tank, then water recirculates. In a direct system, a water source is used and no pump or tank is necessary.

All irrigation systems are usually made up of other basic devices such as electricity connections, control unit, solenoid valves, mesh filter, distribution pipe, and drip emitters. Furthermore, in hydroponic systems, plant nutrition is delivered by a fertilizer injection system that takes nutrients from a tank or reservoir and releases controlled doses of fertilizer into the irrigation system (fertigation). Macronutrients, such as nitrogen (N), to promote vegetative growth, phosphorus (P_2O_5), to contribute to healthy roots and flower blooms, and potassium (K_2O), important to fight off disease and resistance to pests, are given together with meso- and micronutrients responsible for cell wall development, nitrogen and sugar metabolism, and protein synthesis. These systems require more maintenance as the pH, the conductivity, the water salinity, and the total dissolved solids must be monitored and adjusted when necessary. The most sophisticated irrigation systems include remote control facilities of these parameters (Blanc 2012). The natural water retention properties of some substrate based systems (in particular high quality, water-retentive growing medium) allow irrigation systems to be easy in design and construction. In soil-based systems, a long-term (6–8 months) controlled release fertilizer is mixed with the growing medium as plant starter, then fertigation is used.

In a sustainable perspective, important strategies to guarantee a regular irrigation and, at the same time, to minimize the water consumption could be the recovery of rainwater and the recycling of water captured in the drainage system. Moreover, sensor-based systems that automatically monitor the weather conditions (air humidity, air temperature, wind direction and speed, rainfall, etc.) and modify the moisture parameter of plants are also useful (Manso and Castro-Gomes 2015).

2.2.4 Design, Maintenance, and Cost

With regard to design, it is important to look at the following issues (Jim 2015):

- microclimatic conditions and on-site analysis (exposure, lighting, air humidity, air temperature, wind direction and speed, rainfall quantity, etc.);
- structural factors (wall type, geometry and dimension, wall height, texture surface, windows, etc.);
- policy context (legislation and guidance, building regulations, and standards or other initiatives).

The maintenance of LWS is a key factor in its success and should be provided since the design stage, so that aspects such as safe and easy maintenance access are not neglected. As it is known, once installed, all LWS require regular maintenance to guarantee the optimal agronomical and technological performance.

Maintenance procedures are considered as "routine" and "extraordinary". Routine maintenance provides for the following processes: irrigation equipment check; vegetation growth monitoring; number of pruning planned per year; in-service control of materials and components. Extraordinary maintenance provides for the subsequent processes: irrigation system repair; material and components repair and replacement; treatments against pests and diseases.

Routine maintenance can be furthermore organized through two main levels according to plants care frequency: average frequency (less than or equal to two maintenance processes per year) and high frequency (more than two maintenance processes per year).

A detailed maintenance plan should be developed and periodically reviewed to ensure all maintenance needs and the LWS quality in its expected lifetime.

The initial costs can significantly vary from 400 to 1200 €/m^2, depending on living wall solution and material used, and on design specific features and complexity.

In a life cycle costing, not only the installation cost should be taken into account but also the maintenance and disposal cost, varying according to the system adopted and plants selected (Perini and Rosasco 2013).

2.3 LWS and Ecosystem Services

"*Ecosystem services are the benefits that humans derive from ecosystems. They support directly or indirectly our survival and quality of life*" (Costanza et al. 1997). Healthy ecosystems are needed in order to obtain sustainable cities. They affect human well-being and economic activities (TEEB 2011). Ecosystem services have been regulated at European level by the Millennium Ecosystem Assessment (MEA) in 2005. The MEA aim was to evaluate the consequences of the ecosystem change for human well-being and to find the scientific actions needed to enhance the conservation and sustainable use of those systems. The document outlined the strategies

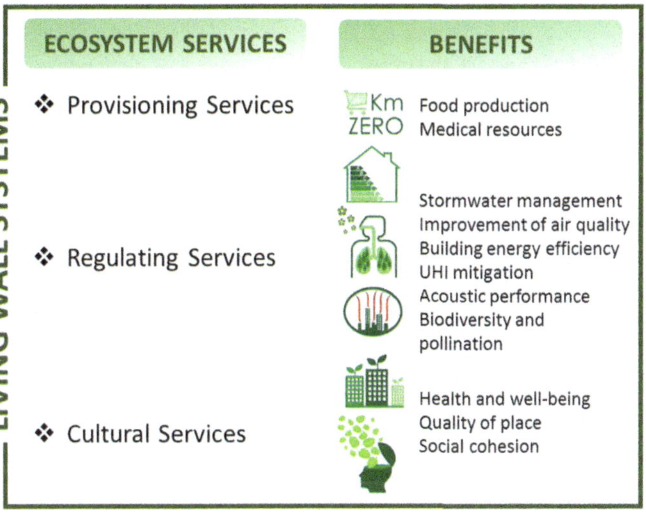

Fig. 2.5 Scheme of the ecosystem services potentially provided by LWS

for the future and showed that the urban context will play in the future an important role for its ecological and environmental functions.

Sometimes to define what is urban, we refer to the density of economic activity, but the criteria are extremely variable (Elmqvist et al. 2013). Urban ecosystem services are generated in different habitats, such as green spaces, including parks, urban forests, cemeteries, courtyards, campus areas, and blue spaces, including courses water, lakes, and ponds (Elmqvist et al. 2015). Green in cities favors a lowering of temperature, reducing the phenomenon of the heat island. Urban greening also helps in regulating air quality by reducing atmosphere pollutants. In the city, also green walls provide important ecosystem services, especially in highly built areas, where green spaces are small and fragmented, for example in industrial buildings, or even on any residential or recreational areas.

Following the MEA categories of ecosystem services, we tried to select the ones potentially provided by LWS (Fig. 2.5) as provisioning, regulating, and cultural services, that are then explained in the following paragraphs.

2.3.1 Provisioning Services

Provisioning services are the material and energy resources that interact, but above all that form, the ecosystem, such as food, raw materials, freshwater, and medical resources. LWS can provide some of them, particularly food and medicinal resources.

Food production

Food production is one of the most important services that ecosystems provide. Cultivation of crops is also taking more and more feet within cities, so that more attention and consideration is required.

There are numerous forms of agriculture in the cities, both traditional and innovative, using innovative techniques and materials, but always trying to safeguard the surrounding environment, for example, saving water in highly populated areas with controlled leaching.

Surely LWS are one of these new technologies useful to produce food and make cities greener (Fig. 2.6). There are numerous experiments on green walls where lettuces, tomatoes, basil, and spinach are grown, and it is also possible to combine some ornamental species with vegetable plants that enhance the aesthetic appearance or take into account the practice of intercropping (Devecchi et al. 2013; Mårtensson et al. 2016).

The possibility to realize these systems on any building's wall increases the available surface for food production, which also has the peculiarity of being km 0, and thus saves energy used to get it up there.

Fig. 2.6 Producing food on a technological living wall proposed by the USA Pavillon at Milan EXPO 2015

Medical resources

In LWS, it is possible to cultivate not only horticultural plants but also herbs such as lavender, melissa, and mint, which are very useful in phytotherapy, pharmaceutical, and cosmetic fields. Cultivation of medicinal herbs is widespread and interest in this kind of culture is growing rapidly.

Of course, this kind of cultivations may be considered at least for self-consuming and more for greening purposes, rather than for eating citizens.

2.3.2 Regulating Services

Stormwater management

Most urban surfaces and built environment are waterproof and do not allow the drainage of the stormwater. Stormwater turns immediately into receiving water bodies. The increase of both runoff volume and peak flow stresses urban stormwater infrastructure. Vegetated areas absorb surplus water, filter, and slowly release it back into the environment, helping the water management in cities (Weinmaster 2009).

International scientific literature evidences the importance of vegetation, such as green roofs, as an urban stormwater management strategy. Researches regarding green wall are extremely limited and statistics on the impact of vertical gardens on stormwater are not currently available. Green walls are able to retain water, in terms of the rainfall interception capacity, thanks to the absorption ability of the substrate, but in a lesser degree than green roofs (Bit 2012). The role of the vegetation in stormwater retention is related to the evaporation and transpiration by plant. The substrate and the other LWS materials should also provide reduction peak of runoff. A system for collection of rainwater, into storage tanks, allows the reuse through fertigation system, according to the quantity necessary for the plant species.

Green walls can also contribute to the improvement of water quality, by removing pollutants from runoff. However, such systems can also release chemical compounds resulting from the use of fertilizers. For this reason, it is good practice to realize systems of collecting and reusing fertigation water. At the same time, these solutions limit the waste of potable water, as the water is returned to the irrigation system and reused for new waterings.

The evidence highlighted above shows that quantifying the LWS stormwater management requires detailed modeling of microclimatic conditions, permeability of surfaces, existing water infrastructure, etc.

Improvement of air quality

Literature recognizes important effects of vegetation on air quality improvement, not only because plants and leaves can absorb carbon dioxide (CO_2) and release oxygen through photosynthesis but also plants can reduce other air pollutants such as nitrogen oxides (NOX), volatile organic compound (VOCs), and dust. Some studies show also the contribution in carbon storage sequestration from soil respiration (plant roots and associated microorganisms) (Pouyat et al. 2006, 2009).

LWS are significant green infrastructures for urban environment, in particular in street canyons, where other types of greenery solutions, as trees, could worsen the street-level pollutant concentration. In a dense built and canyon-like environment, in fact, the aerodynamic influence of trees could outweigh their filtration capacity (Buccolieri et al. 2009; Vos et al. 2013; Wania et al. 2012). On the contrary, the integration of the vegetation in the building envelope does not preclude the ventilation (Perini 2013) and has a positive effect on air quality (Pugh et al. 2012).

LWS contribute to the reduction of air pollutants even indirectly, in accordance with the best building energy performance and lower consumption for heating and, especially, cooling, resulting in the reduction of greenhouse gases emissions (see below).

According to Feng and Hewage (2014), the air cleaning potential of living walls should be assessed in the whole lifecycle, from manufacturing to disposal, and it is influenced by the materials and plants chosen for the system. Researchers compare two types of LWS as defined in this chapter, a continuous felt layer system and a modular HDPE panel system. Their analysis shows that the first releases three times more toxic substances to the environment than the second one. Additionally, the felt layer system needs as many as 23 years to balance the emissions and cannot offset the pollution it initially created during its expected operating life, amounting to 10 years. Therefore, among the typical LWS evaluated, the felt layer system is the less environmentally friendly solution, in terms of the air pollution abatement.

If the improvement in air quality is a desired performance, plant species have to be carefully selected for their ability to capture dust or toxins and, in particular polluted environments (e.g., next to a busy road or a construction site) for their longevity and robustness.

Finally, according to Perini et al., the capacity of living wall systems to remove the pollutants is equal to 50% of that of a green roof with the same plants and the same surface. Researchers correlate also the air quality improvement to the public health costs reduction.

Buildings energy efficiency

LWS allows to improve the buildings energy performance during cooling and heating seasons, reducing greenhouses gas emission, as a relevant number of studies, dating back to end of 80s, has clearly demonstrated. Nevertheless, given the complex behavior of LWS and the number of variables involved, still a lack of generalized data for architects and designer is observed in literature.

Vegetated walls actually constitute a dynamic system, whose behavior is quite difficult to be modeled in simulation tools due to the complexity of physical phenomena involved and to the lack of proper input data. Experimental characterization can represent a viable and reliable way to collect information since the living walls are exposed to real boundary conditions. Data collected through long-term monitoring campaigns are nevertheless not simple to be compared. Results are influenced by the climate and weather conditions, the wall assembly, the wall orientation, plants typology, and health conditions.

In a recently published review on LWS (Coma et al. 2017), it is highlighted that the efficiency of this system to save energy is strictly related to the technology construction choice, the climatic influence, the type of plants chosen, and finally the synergy between the different elements of the wall.

During the winter season, LWS act in two different ways on the wall performance: the leaves and the plants are able to considerably reduce the wind speed in the vicinity of the wall with a consequent improvement of the external thermal surface resistance of the wall (Perini et al. 2011). Second, the LWS, when applied on a vertical façade, create an air layer between the rears of the vegetated module and the wall surface. This air gap (usually naturally ventilated) is an extra additional layer for the wall because the air has a higher temperature than outer air. Being the wall in contact with a cavity with warmer air temperature than the external one, the heat losses through the wall are reduced (Bianco et al. 2016).

During summer, the positive effects are given by the presence of the green layer which reduces the solar heat gain transmitted through the building envelope. Leaves shading effect can be considered as a drawback during the winter period, for this reason, some studies propose the use of non-evergreen species or to integrate the LWS on a non-south façade. The plants use the solar radiation for their biological activities, such as evapotranspiration and photosynthesis, and they also store a quota of solar radiation by the lymphatic system. A non-secondary effect to reduce the surface temperature of the wall in summer is the evapotranspiration effect of plants. For these two mechanisms, it is possible to observe a lower surface temperature on the LWS when compared to a reference non-vegetated technology with same exposition. Literature shows that LWS is able to reduce the external surface temperature in a wide range between 1 °C and 31.9 °C (Coma et al. 2017).

Urban heat island mitigation

The benefits deriving by the ability of LWS to reduce the external surface temperature are also reflected on the mitigation of UHI. This effect is particularly evident in narrow streets, surrounded by buildings which create a canyon effect, with low air circulation and a consequent increase of the outdoor air temperature. LWS would improve the inhabitant thermal comfort by means of leaves shading, reflection coefficient decreasing, evapotranspirative cooling and direct, and diffuse solar radiation attenuation (Alexandri and Jones 2008). On the subject of the urban scale, only few studies are presented in literature and mainly related to modeling activity, but a lack of monitoring campaigns carried out on extensive urban areas vegetated with LWS is observed. A simulation study (validates against experimental data of the LWS) in La Rochelle (France) showed an air temperature reduction of 1.5 °C in street canyon with LWS if compared to non-vegetated canyon (Djedjig et al. 2016). In order to assess these phenomena, an experimental study in Torino (north Italy) analyzed the external surface temperatures of three different tested claddings on a demonstration building, one with LWS, one plastered, and one wood finished. The temperatures measured during summer on LWS are significantly lower than those measured on the other claddings and more constant, when solar radiation is present. In particular, the peak surface temperature difference between LWS and the plastered wall (south façade) reaches the value of 11 °C (Serra et al. 2014).

Acoustic performance

Also for the acoustic aspects, as for the energy-related performance, the benefits of the vegetation play a double role on the building and on the urban environment as well. Acoustic performance of vegetated walls is improved thanks to an increased sound insulation and sound absorption given by the presence of the plants as well as of the substrate. Urban level LWS, when applied on buildings on a large scale, can efficiently reduce the propagation of low and middle frequencies (Pérez et al. 2016). These results were highlighted also by Wong et al. in 2010, whose study demonstrated the absorbing effect of the substrate for these frequencies and a lower impact on high frequency, due to the scattering effect.

Starting from a laboratory experience, a Spanish research demonstrated that a green wall could strongly improve the acoustic performance of a wall. A weighted sound reduction index (R_w) of 15 dB and a weighted sound absorption coefficient (α) of 0.40 was measured (Azkorra et al. 2015). Moving from laboratory to in situ measurements, the results were less promising. The measurement carried out in Spain, Leida showed that a layer of vegetation (20–30 cm) positioned on a vertical wall was able to improve the sound insulation of 1 dB for traffic noise and between 2 and 3 dB for pink noise.

A laboratory investigation carried out by Serra et al. (2017) showed that the most important role in terms of sound absorption is due to the substrate principally, rather than to the vegetation. The measured values of sound absorption are considered of relevant importance if compared to the traffic noise level typical spectra.

Biodiversity and Pollination

LWS, as part of the green infrastructure, are also involved in the urban ecosystem contributing at improving plants, insects, and birds biodiversity. In particular, the living wall can provide habitat and food for bees, supporting native pollinator communities within the city. No specific research on this topic is yet available.

2.3.3 Cultural Services

The city is an environment full of stimuli, possibilities, but it is also stressful for those who live there that often have an extremely lively lifestyle with a limited time to reflect and contemplate. The recreational aspects of urban greening are very important in cities, together with the aesthetic and cultural values (Bolund and Hunhammar 1999).

As defined by MEA, the cultural services are the nonmaterial benefits that people derive from ecosystems including: cultural diversity; spiritual and religious values; educational values; inspiration; aesthetic values; social relations; sense of place; values of cultural heritage; recreation and ecotourism.

Perceptions of cultural services are more likely to differ between individuals and communities than, for example, perceptions about the importance of food production (MEA 2003). The cultural ecosystem services bring the different values and meanings that people find in nature to increase their awareness and consciousness about the environment, with a look on sustainable development. The integration of the green in the built environments has an important role for the psychological well-being, quality of place, and social cohesion. Living walls can help to address the lack of green space in urban areas.

Health and well-being

The literature shows as nature has therapeutic and regenerative effects. There is considerable scientific evidence on health-related benefits got by hospitalized patients by looking green areas. Those benefits are a combination of positive changes both physiological (for instance in blood pressure, heart activity) and psychological/emotional (stress reduction, positive feelings). Furthermore, the clinical outcomes show that these patients have shorter recovery time and they suffered fewer complications. Overall costs healthcare saving are expected and often proved (Ulrich et al. 2008). Others studies (Ulrich 1981; Beatley 2011; White and Gatersleben 2011) suggest also that people generally prefer a view of natural settings rather than built environments (rooms, buildings, town), because it is related to recovery or restoration from stress. In the workplace, for instance, the nature experience improves employees' satisfaction, enthusiasm, and concentration, it increases the ability to generate creative ideas, and it reduces frustration and absenteeism as well (Bringslimark et al. 2009).

Other effects on health and well-being of exposure to nature are well documented. According to the World Health Organization, stress is a common cause of both mental health disorders and cardiovascular diseases. Living walls should have a significant effect on stress reduction, allowing a direct access to nature.

Quality of place

As perfectly explained by James (1892), fascination is a kind of involuntary attention, that is the attention that does not require mental effort but is attracted by stimuli that have fascinating qualities directly. Most sensations are signs for us of other sensations whose space-value is believed to be more real. Over one century ago, the author assumed that nature has the advantage over the built, being full of objects and processes that are not challenging but fascinating. In designing or transforming an urban area, there are often many constraints and conditions to consider. Space is a scarce resource in many social and urban contexts, as it is subjected to different uses, as there are numerous requests for its use. For these reasons in these areas, the realization of LWS can be a solution to solve this problem. It is good to remember that design ideas have to be approved from a technical and economic point of view and that it is necessary to predict who will benefit the area in order to make it accessible to everyone. In addition, it is essential to understand what will be the maintenance tasks involved. Different aspects of restoration (fascination, distance, etc.) can be used to increase the size of restorativeness and the intended use of a particular environment (Ivarsson and Hagerhall 2008). The vegetation is aesthetically pleasing and can promote restoration (Dunnett and Kingsbury 2004), in this case, the aesthetic value of the LWS is due to their greater visibility. A comparative analysis between the amount of vertical and horizontal surfaces in urban environments showed that the availability of greening vertical surfaces is always equal if not greater than the horizontal ones (Bellomo 2003).

Social cohesion

The integration of vegetation in the urban environment can improve community cohesion through the networks of formal and informal relationships among neighborhood residents (Sullivan et al. 2004). It can also even reduce violent conduct and crime (Kuo and Sullivan 2001). Moreover, the presence of green spaces in the urban environment induces people to use outdoor spaces for social activity (Sullivan et al. 2004).

The LWS may cover collective spaces, based on volunteer work, and the offer to community members the chance to get in direct contact together. New networks and community ties contribute to the physical and social regeneration of urban areas. An example of this approach is provided by the Los Angeles Urban Farming Food Chain project, where a public vertical gardening system offers direct access to fresh products, greens the environment, and provides an opportunity for community service and involvement. In addition, the vertical greenery can have an important educational and cultural value. For example, it may promote the knowledge of botany and the natural sciences among the community. Living walls are installed in some schools and specific educational programs for students are offered, for instance, to learn about the seed-to-harvest cycle (e.g., vertical farm).

2.4 Conclusions and Perspectives

LWS are usually associated with green roofs for their application on buildings. While green roofs are an established technology in the construction sector, only recently many designers are increasing the interest in vertical greening systems.

As explained in the previous paragraphs, LWS can provide several services both to the urban environment and at the building scale. Otherwise, some limits to their development are reported, such as the high initial and maintenances costs; the lack of scientific knowledge; the absence of economic incentives as well as of technical standards.

In Europe (Germany, UK, and France), different programs, action plans, and guidelines are now available. An Italian technical standard concerning the green roof was adopted (UNI 11235:2015) and an Italian standard for LWS may be proposed (Giordano et al. 2017).

Socioeconomic barriers (maintenance, negative perception, mechanism/policy) can be overcome improving the knowledge on LWS and their positive role in the urban ecosystem. In particular, assessing the potential benefits provided by living wall in terms of cultural services is an open issue.

Studies on irrigation-free living walls and the use of xerophyte plants are under consideration.

In conclusion, sustainable LWS that provide architectural and environmental benefits through the use of independent modular units, renewable materials, environmental friendly substrates, and native plants is the key for their future development.

References

Alexandri E, Jones P (2008) Temperature decreases in a urban canyon due to green walls and green roofs in diverse climate. Build Environ 43(4):480–493

Azkorra Z, Pérez G, Coma J, Cabeza LF, Bures S, Alvaro JE, Erkoreka A, Urrestarazu M (2015) Evaluation of green walls as a passive acoustic insulation system for buildings. Appl Acoust 89:46–56

Beatley T (2011) Biophillic cities. Integrating nature into urban design and planning. Island Press

Bellomo A (2003) Pareti verdi-Linee guida alla progettazione. SE, Napoli

Benvenuti S, Malandrin V, Pardossi A (2016) Germination ecology of wild living walls for sustainable vertical garden in urban environment. Sci Hortic 203:185–191

Bianco L, Serra V, Larcher F, Perino M (2016) Thermal behaviour assessment of a novel vertical greenery module system: first results of a long-term monitoring campaign in an outdoor test cell. Energ Effi 1–14

Bit E (2012) L'approccio idrico al sistema vegetale: il sistema automatizzato d'irrigazione. In: Il nuovo verde verticale. Tecnologie, progetti, linee guida. Wolters Kluwer Italia, Milano, p 243

Blanc P (2012) The vertical garden: from nature to the city. W. W. Norton & Company, New York

Bolund P, Hunhammar S (1999) Ecosystem services in urban areas. Ecol Econ 29:293–301

Bringslimark T, Hartig T, Patil GG (2009) Psychological benefits of indoor plants in workplaces: putting experimental results into context. HortScience 42:581–587

Buccolieri R, Gromke C, Di Sabatino S, Ruck B (2009) Aerodynamic effects of trees on pollutant concentration in street canyons. Sci Total Environ 407:5247–5256

Coma J, Pérez G, de Gracia A, Burés S, Urrestarazu M, Cabeza FL (2017) Vertical greenery systems for energy savings in buildings: a comparative study between green walls and green facades. Build Environ 111:228–237

Costanza R, d'Arge R, de Groot RS, Farber S, Grasso M, Hannon B, Limburg K, Naeem S, O'Neill RV, Paruelo J, Raskin RG, Sutton P, van den Belt M (1997) The value of the world's ecosystem services and natural capital. Nature 387:253–260

Devecchi M, Merlo F, Vigetti A, Larcher F (2013) The cultivation of mediterranean aromatic plants on green walls. Acta Hortic 999:243–247

Djedjig R, Bozonnet E, Belarbi R (2016) Modeling green wall interactions with street canyons for building energy simulation in urban context. Urban Clim 16:75–85

Dunnett N, Kingsbury N (2004) Planting green roofs and living walls. Timber Press, Portland

Elmqvist T, Fragkias M, Goodness J, Güneralp B, Marcotullio PJ, McDonald RI, Parnell S, Schewenius M, Sendstad M, Seto KC, Wilkinson C (2013) Urbanization, biodiversity and ecosystem services: challenges and opportunities. A global assessment. Springer, Dordrecht, Heidelberg, New York, London, pp 3–4

Elmqvist T, Setälä H, Handel SN, van der Ploeg S, Aronson J, Blignaut JN, Gómez-Baggethun Nowak EDJ, Kronenberg J, de Groot R (2015) Benefits of restoring ecosystem services in urban areas. Curr Opin Environ Sustain 14:101–108

Feng H, Hewage K (2014) Lifecycle assessment of living walls: air purification and energy performance. J Clean Prod 69:91–99
Francis RA, Chadwick MA (2012) What makes a species synurbic? Appl Geogr 32:514–521
Francis RA, Lorimer J (2011) Urban reconciliation ecology: the potential of living roofs and walls. J Environ Manage 92:1429–1437
Giordano R, Montacchini E, Tedesco S, Perone A (2017) Living wall systems: a technical standard proposal. Energy Procedia 111:298–307
Ivarsson CT, Hagerhall CM (2008) The perceived restorativeness of gardens—assessing the restorativeness of a mixed built and natural scene type. Urban For Urban Greening 7(2):107–118
James W (1892) Psychology: the briefer course. Henry Holt & Company, New York
Jim CY (2015) Greenwall classification and critical design-management assessments. Ecol Eng 77:348–362
Kuo FE, Sullivan WC (2001) Environment and crime in the inner city does vegetation reduce crime? Environ Behav 33(3):343–367
Larcher F, Scariot V (2009) Assessment of partial peat substitutes for the production of *Camellia japonica*. HortScience 44(2):312–316
Loh S (2008) Living walls—a way to green the built environment. BEDP Environ Des Guide, 1(TEC 26) 1–7
López-Cuadrado MC, Ruiz-Fernández J, Masaguer A (2006) Producción De Planta Ornamental En Contenedor Con Sustratos Alternativos a La Turba. IMIDRA, Madrid Espa˜na
López-Rodríguez G, Pérez-Esteban J, Ruiz-Fernández J, Masaguer A (2016) Behavior and evolution of sustainable organic substrates in a vertical garden. Ecol Eng 93:129–134
Manso M, Castro-Gomes J (2015) Green wall systems: a review of their characteristics. Renew Sustain Energy Rev 41:863–871
Mårtensson L-M, Wuolo A, Fransson A-M, Emilsson T (2014) Plant performance in living wall systems in the Scandinavian climate. Ecol Eng 71:610–614
Mårtensson L-M, Fransson A-M, Emilsson T (2016) Exploring the use of edible and evergreen perennials in living wall systems in the Scandinavian climate. Urban For Urban Greening 15:84–88
Masaguer-Rodríguez A, López-Fabal A, Carmona-Chiara E, Fornés-Sebastiá F, Ordovás-Ascaso J, Gómez-Sánchez MA, Moreno-Aguirre MT, Marfá-Pagès O, Cáceres-Reyes R, López-Nú˜ nez R, Belda R (2015) Uso del compost como componente de sustratos para cultivo en contenedor. Tomo 2. Recursos Orgánicos: Aspectos Agronómicos Y Medioambientales, vol III. Mundi Prensa, Espa˜ na, p 244
Millennium Ecosystem Assessment (2003) Ecosystems and human well-being: a framework for assessment. Island Press, Washington, pp 71–83
Ottelé M, Perini K, Fraaij ALA, Haas EM, Raiteri R (2011) Comparative life cycle analysis for green façades and living wall systems. Energy and Build 43(12):3419–3429
Pérez G, Coma J, Martorell I, Cabeza LF (2014) Vertical greenery systems (VGS) for energy saving in buildings: a review. Renew Sustain Energy Rev 39:139–165
Pérez G, Coma J, Barreneche C, de Gracia A, Urrestarazu M, Burés S, Cabeza LF (2016) Acoustic insulation capacity of vertical greenery systems for buildings. Appl Acoust 110:218–226
Perini K (2013) Progettare il verde in città. Franco Angeli, Milano
Perini K, Rosasco P (2013) Cost-benefit analysis for green façades and living wall systems. Build Environ 70:110–121
Perini K, Ottelé M, Fraaij ALA, Haas EM, Raiteri R (2011) Vertical greening systems and the effect on air flow and temperature on the building envelope. Build Environ 46(11):2287–2294
Pouyat RV, Yesilonis ID, Golubiewski N (2009) A comparison of soil organic carbon stocks between residential turfgrass and native soil. Urban Ecosyst 12:45–62
Pouyat RV, Yesilonis ID, Nowak DJ (2006) Carbon storage by urban soils in the United States. J Environ Qual 35:1566–1575
Pugh TAM, MacKenzie AR, Whyatt JD, Hewitt CN (2012) Effectiveness of green infrastructure for improvement of air quality in urban street canyons. Environ Sci Technol 46(14):7692–7699

Serra V, Candelari E, Giordano R (2014) Vertical greening systems and urban heat island related aspects: outcomes of a research project. In: Third international conference on countermeasures to urban heat island, Venezia, 13–15 ottobre 2014, pp 1410–1421
Serra V, Bianco L, Candelari E, Giordano R, Montacchini E, Tedesco S, Larcher F, Schiavi A (2017) A novel vertical greenery module system for building envelopes: The results and outcomes of a multidisciplinary research project. Energy Build 146:333–352
Shaikh AF, Gunjal PK, Chaple NV (2015) A review on green walls technology, benefits and design. Int J Eng Sci Res Technol 4(4)
Sheweka SM, Mohamed NM (2012) Green Facades as a new sustainable approach towards climate change. Energy Procedia 18:507–520
Sullivan WC, Kuo FE, Depooter SF (2004) The fruit of urban nature vital neighborhood spaces. Environ Behav 36(5):678–700
TEEB—The Economics of Ecosystems and Biodiversity (2011) TEEB manual for cities: ecosystem services in urban management
Ulrich RS (1981) Natural versus urban scenes: some psychophysiological effects. Environ Behav 13:523–556
Ulrich RS, Zimring C, Zhu X, Du Bose J, Seo HB, Choi YS, Quan X, Anjali J (2008) A review of the research literature on evidence-based healthcare design. Health Environ Res Des 1:61–125
Vos PEJ, Maiheu B, Vankerkom J, Janssen S (2013) Improving local air quality in cities: to tree or not to tree? Environ Pollut 183:113–122
Wania A, Bruse M, Blond N, Weber C (2012) Analysing the influence of different street vegetation on traffic-induced particle dispersion using microscale simulations. J Environ Manage 94:91–101
Weinmaster M (2009) Are green wall as green as they look? An introduction to the various technologies and ecological benefits of green wall. J Green Build 4(4):3–5
White EV, Gatersleben B (2011) Greenery on residential buildings: does it affect preferences and perceptions of beauty? J Environ Psychol 31(1):89–98

Chapter 3
Improving the Metabolism and Sustainability of Buildings and Cities Through Integrated Rooftop Greenhouses (i-RTG)

David Sanjuan-Delmás, Pere Llorach-Massana, Ana Nadal, Esther Sanyé-Mengual, Anna Petit-Boix, Mireia Ercilla-Montserrat, Eva Cuerva, Maria Rosa Rovira, Alejandro Josa, Pere Muñoz, Juan Ignacio Montero, Xavier Gabarrell, Joan Rieradevall and Oriol Pons

Abstract Food security in cities is an increasing concern due to the impact of climate change and the concentration of world population in cities. Urban agriculture (UA) aims at enhancing food production in urban areas, providing potential environmental

---

D. Sanjuan-Delmás · P. Llorach-Massana · A. Nadal · A. Petit-Boix · M. Ercilla-Montserrat
M. R. Rovira · X. Gabarrell · J. Rieradevall
Sostenipra Research Group (SGR 01412), Institute of Environmental Sciences and Technology (MDM-2015-0552), Z Building, Universitat Autònoma de Barcelona (UAB), Campus UAB, 08193 Bellaterra, Barcelona, Spain

E. Sanyé-Mengual
Research Centre in the Urban Environment for Agriculture and Biodiversity (ResCUE-AB), Alma Mater Studiorum University of Bologna, Viale Giuseppe Fanin, 42-40127 Bologna, Italy

E. Cuerva
Department of Construction Engineering, School of Industrial Engineering (ETSEIB), Universitat Politècnica de Catalunya (UPC-Barcelona Tech), Avinguda Diagonal 647, 08028 Barcelona, Spain

A. Josa
Department of Civil and Environmental Engineering, School of Civil Engineering (ETSECCPB), Sustainability Institut (IS.UPC). Universitat Politècnica de Catalunya (UPC-Barcelona Tech), Jordi Girona 1-3, 08034 Barcelona, Spain

A. Josa
Institute for Sustainability Science and Technology (IS.UPC), Universitat Politècnica de Catalunya (UPC-BarcelonaTech), Campus Nord, C/Jordi Girona 31, 08034 Barcelona, Spain

P. Muñoz · J. I. Montero
Institute of Food and Agricultural Research (IRTA), Carretera de Cabrils, Km 2, 08348 Barcelona, Spain

X. Gabarrell · J. Rieradevall
Department of Chemical, Bilogical and Environmental Engineering, Universitat Autònoma de Barcelona (UAB), Campus UAB, 08193 Bellaterra, Barcelona, Spain

O. Pons (✉)
Department of Architectural Technology, School of Architecture (ETSAB), Universitat Politècnica de Catalunya (UPC-Barcelona Tech), Avinguda Diagonal 649, 08028 Barcelona, Spain
e-mail: oriol.pons@upc.edu

D. Nandwani (ed.), *Urban Horticulture*, Sustainable Development and Biodiversity 18, https://doi.org/10.1007/978-3-319-67017-1_3

advantages by reducing food transport, packaging and waste generation. Among UA alternatives, rooftop greenhouses (RTGs) are greenhouses built on top of urban roofs, in which mainly soil-less agriculture systems are used to produce food. When RTGs are integrated into the metabolism of their buildings, they exchange CO_2, energy and water to improve their performance. This alternative is called integrated RTG (i-RTG). This chapter analyses the use of i-RTGs to improve buildings and cities' metabolism and its particular application in the area of Barcelona. This analysis aims to define a new agricultural system from a technological and sustainability approach focusing on Mediterranean cities. Our research is based on the development and results of the Fertilecity project. A particular experimental analysis was conducted at ICTA's i-RTG lab located near Barcelona. The main factors of interest are architectural and engineering requirements, urban integration, CO_2 emissions management, energy consumption, food production, social integration and rainwater harvesting. This analysis has used different methods such as life cycle assessment (LCA), life cycle costing (LCC) and semi-quantitative assessments. Multiple integrated results were obtained both at the building and city scale. For example, we proved that the i-RTG and its flow exchanges with the building could help to save heating energy, waste generation, water consumption and CO_2 emissions.

Keywords Food security · Urban agriculture · LCA
Building metabolism · Circular economy · Industrial ecology

3.1 Introduction

Cities are currently known to be the habitat of most of the world's population, which is estimated to augment up to 66% by 2050 (United Nations 2014). For example, 75% of the people of Europe lives in urban centres, and this percentage is predicted to reach 80% in 2020 (European Commission 2010). Similarly, 79% of the Latin American and Caribbean population live in urban settlements (da Cunha and Rodríguez-Vignoli 2009). In this sense, metropolises have to deal with the demand for residential, educational and work facilities, among others, whereas research aims to meet the dwellers' aspirations (Joffe and Smith 2016) and help cities become smart and sustainable (Lugaric and Krajcar 2016). Nevertheless, cities are not self-sufficient regarding energy, water and food provision, which are daily basic needs. Furthermore, recent research studies and events proved that cities are more prone to natural disasters associated with climate change, and their negative effects worsen due to high-density urban population (Deely et al. 2010). Urban daily needs and the consequences of climate change hazards stress the need to innovate our cities to meet these demands in a smart and sustainable way. For instance, urban food provisioning relies on remote agricultural areas (dos Santos 2016). In consequence, urban food security depends on these remote production areas and its associated transportation and distribution costs and impacts, as well as its stability, variety and quality. In this sense, recent research projects study smart cities that produce their food, which improves their environmental behaviour by cutting down food transport, packaging and waste generation, among others (Pons et al. 2015; dos Santos 2016).

Local food production is not a new concept, as ancient metropolises already produced agricultural goods from a self-sufficiency perspective. For instance, major settlements in ancient (e.g. Roman, Greek or Byzantine) and Middle Age civilisations cultivated vegetables and herbs in gardens and orchards (Janick 2002; Zaar 2011). Until the present time, different forms of urban agriculture (UA) have been developed in cities, especially during critical periods. For instance, during World War I and II, urban allotments were an important food resource given its limited availability (Spudic 2007). Moreover, other initiatives provided social support, such as the 'City Farms and Community Gardens' project that started in the Netherlands and UK in the 1970s (McKay 2011). Some of the most recent UA experiences and projects focus on smart and sustainable cities (Pons et al. 2015; dos Santos 2016), zero emissions (Arredondo 2013) and green design. At present, UA initiatives could provide environmental and cost improvements, contribute to the food security of the vulnerable population and change the established linear food production process through an optimised, circular agro-urban system (Zaar 2011).

Vertical farming (VF) is a visionary form of UA (Morán and Aja 2011), which is made up of growing food in and around buildings. Previous literature has stated different concepts within the framework of vertical farming, such as skyfarming (Spudic 2007) and zero-acreage farming (Chalana 2014). Rooftop greenhouses (RTGs) are located on the top of buildings and represent an innovative alternative because they add more agricultural areas to the limited green spaces in urban areas. They also promote food self-sufficiency and take advantage of underused edifices roofs in the city. The advantages above are of growing interest to the city, where local authorities and citizens are rethinking their roofs potential applications to obtain further benefit (Contreras and Castillo 2015).

RTGs are assembled on flat roofs and are made up of horticultural commercial holdings, but there are also cases of public, self-consumption and research RTGs that focus on issues such as education, nutrition, recreation and sustainability (Nadal Fuentes 2015). Although some RTGs follow traditional agriculture techniques, the majority of these greenhouses use advanced cultivation methods, such as the soil-less hydroponic or aeroponic systems (Lakkireddy 2012). Most RTGs grow vegetables, but some grow aromatic plants and fruit trees. Additionally, RTGs avoid soil pollution and its related accumulation of metals, which is a common risk in both conventional and urban agriculture (Mitchell et al. 2014). Moreover, the control of trace metals from air pollution is easier because it depends on the location and the distance to traffic areas (Säumel et al. 2012).

A substantial percentage of RTGs concentrate in North American cities, such as the Vinegar Factory in New York (The Vinegar Factory 2017). This building holds a marketplace with an RTG of 830 m^2 that sells vegetables and fruits grown on site. Another example is Lufa Farms, in Montreal (Canada), which has a 2900 m^2 RTG built on a commercial building roof. The production, which includes lettuce, tomatoes and eggplants, is commercialised using a Community Supported Agriculture (CSA) strategy (Lufa Farm 2017). These vegetables are cultivated in several thermal conditioned areas that are controlled using environment technologies. In contrast, RTGs are still experimental experiences in Europe, with examples such as the German INFarming (Fraunhofer UMSICHT 2011).

In the future, RTGs are expected to provide more environmental advantages once they are included in the building design from the very first draft, and their design and construction are optimised according to the edifice properties. Integrating a RTG within the metabolism of the edifice through a symbiosis involving the flows of CO_2, energy, water and food is a solution called integrated RTG (i-RTG), which improves the environmental performance of both systems. The application of air from occupied areas of the building contains extra CO_2 from breathing that can benefit plants for the photosynthesis. Regarding water, harvesting rainwater from the rooftop can provide alternative water sources for the i-RTG. Finally, heating gains and losses from and to the RTG due to its greenhouse effect and ventilation can be used to control the temperature of the greenhouse and improve the heating comfort of the building interiors.

Barcelona is an example of a Mediterranean metropolis. Its climate has a singular soft temperature and average humidity. It has a population of 1.6 million inhabitants and a surface of 102.2 km^2, which defines a population density of 15,706 hab/km^2 (Ajuntament de Barcelona 2016). Almost all buildings have flat roofs that are capable of incorporating an RTG. In Barcelona, UA experiences have been progressively but slowly implemented (Cerón-Palma et al. 2012), with growing initiatives of public and private urban gardens from the 1990s. Moreover, roofs and vertical farming both in educational and residential edifices have been started in recent years, although they are still anecdotal. However, Barcelona has a high potential for implementing RTGs in its unproductive flat roofs, increasing multifunctionality and integrating food production into buildings. Furthermore, implementing i-RTGs could provide solutions to environmental problems by reducing urban CO_2 emissions, water and energy consumption and food imports.

The present chapter focuses on i-RTGs and their capacity to improve buildings and cities' metabolism with a specific case study from the Mediterranean urban area of Barcelona. Its main objective is to define a new agricultural system to be implemented in Mediterranean cities in the future. With this purpose, the following sections describe the novel approach of the Fertilecity project (CTM2013-47067-C2-1-R) in its quest to discover the potential sustainability benefits of i-RTGs, including environmental, economic and social issues. This chapter is based on 3 years of research funded by the Spanish Ministry of Economy and Competitiveness that will continue in the future through Fertilecity II (CTM2016-75772-C3-1-R) (MINECO/FEDER, UE).

3.2 Scope of the Project

The Fertilecity project is short for 'Agro-urban sustainability through rooftop greenhouses. Eco-innovation on residual flows of energy, water and CO_2 for food production'. The main motivation of this research was to assess the potential implementation of RTGs in urban centres and the potential benefits that arise when providing cities with alternative food resources that increase their resilience and sustainability. In this sense, two approaches were considered, i.e. at the building and urban scale (Fig. 3.1). The building scale analysis was based on the i-RTG Lab located in a singular research building (see Sect. 3). This study aims to determine the economic and environmental performance of the i-RTG through experimental data. The issues of interest considered were the infrastructure and the water, energy, food, biomass and CO_2 stocks and flows. An additional urban scale analysis was conducted to understand the requirements of the existing and future building stock when hosting an RTG or i-RTG. The social perception of the main stakeholders was reviewed to assess their acceptance of this new form of agriculture. In the end, a complete sustainability assessment integrated the environmental, economic and social dimensions of this system.

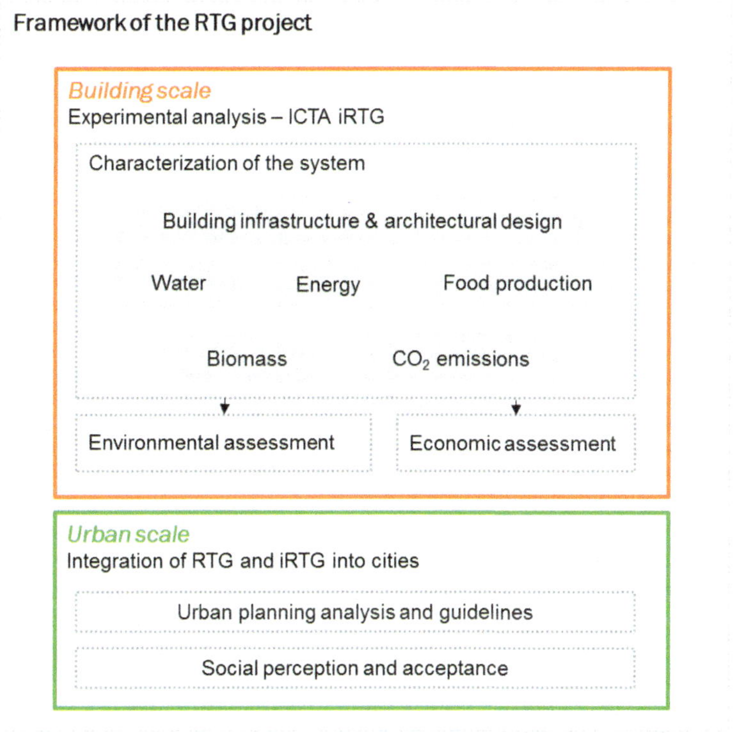

Fig. 3.1 Scope of the RTG project and items under analysis

3.3 System Under Study: The ICTA-ICP Building and the i-RTG

The i-RTG analysed as a case study was built on the ICTA-ICP building, located in the campus of Universitat Autònoma de Barcelona (Spain—Latitude 41° 29′ 51.6″N; Longitude 2° 06′ 31.9″E). The ICTA-ICP building (HA Architects 2011) is a five-storey construction that was designed with the most demanding sustainability criteria: renewable materials, passive heating systems, energy efficiency, multifunctionality, modularity or building-integrated agriculture. Thanks to these criteria, the building was awarded the LEED-Gold® certification. The i-RTG was set up during the construction of the building itself. From an architectural standpoint, it is complex to distinguish the i-RTG from the rest of the building. The main characteristic of the i-RTG is its potential to exchange multiple flows with the ICTA-ICP building (Pons et al. 2015; Nadal et al. 2017) (see Fig. 3.2). These include:

(1) the rainwater harvested by the building, which can be used to irrigate the crops or for other non-potable purposes;
(2) the thermal inertia of the building and the residua air from the ventilation of indoor spaces, which can be used to warm the crops at night; and
(3) the CO_2 generated by human respiration in indoor spaces, which can be injected into the i-RTG for carbon enrichment.

These flow exchanges are of great interest to increase both the building and the i-RTG efficiency. The i-RTG has an area of 122 m^2, of which 84.3 m^2 are devoted to food

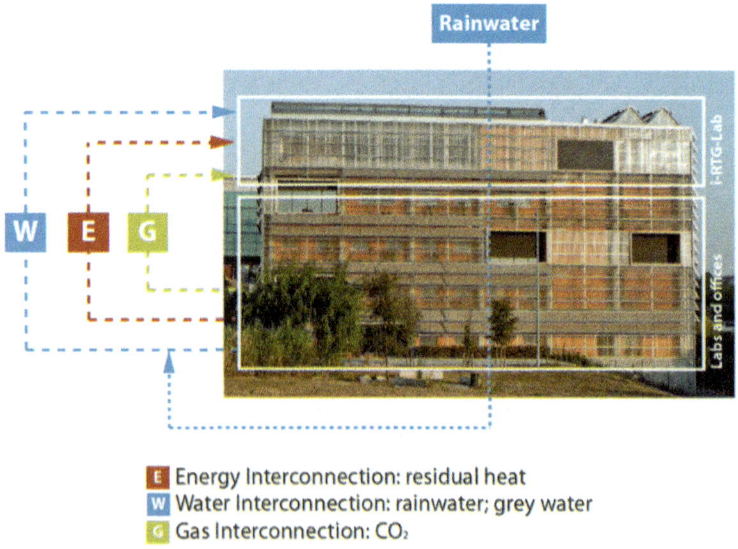

Fig. 3.2 ICTA-ICP building: flow exchanges between the building and the i-RTG

production. In this area, 57 perlite bags (1 m long and 0.35 m diameter, approximate total volume of 40 L) are distributed in 12 lines with a 1.2 m row separation.

3.4 Methodological Framework

The project integrated a set of multi- and interdisciplinary tools to holistically determine the potential benefits and limitations of RTG and i-RTG systems. These are summarised in Table 3.1 based on the scale and item under analysis. The experimental phase was conducted at the i-RTG, and real data were obtained through analytical instruments, such as temperature and CO_2 sensors. The environmental and economic impacts were evaluated through the life cycle assessment (LCA) (ISO 14040:2006; ISO 14044:2006) and life cycle costing (LCC) (ISO 15686-5:2008) methodologies, respectively. At the urban scale, social tools, such as semi-structured interviews, where used to determine user perceptions. Geographic Information Systems (GIS) were needed to identify the best areas of implementation and elaborate normalised protocols. A description of each item under analysis is presented in the following sections.

3.5 Characterization of the System

3.5.1 Infrastructure: i-RTG Requirements

The implementation of the i-RTG was a multidisciplinary project with the participation of architects, engineers, environmental scientists, agronomists, the users of the building and the administration, considering a list of sustainability requirements for the new edifice. The design of the i-RTG had to comply with the current legislation, and the experts had to deal with the maximisation of the architectural and agricultural functions (Sanyé-Mengual et al. 2014b).

The ICTA-ICP building, which holds the i-RTG, is leaning on a four-storey hyperstatic heavy structure composed of poured on site reinforced concrete columns and bidirectional solid slabs. In contrast, the i-RTG has a light isostatic steel skeleton consisting of rectangular hollow section supports, hollow and rolled section trusses—such as $150 \times 150 \times 5$ and $120 \times 200 \times 4$ with UPN 140, respectively—stabilised with bracing composed of steel wire crosses (HA architects 2011). The covering is a horizontal casement with mobile polycarbonate panels that open and close based on the indoor and outdoor conditions for climatization.

The skeleton and the covering have significant differences compared with multi-tunnel greenhouses for conventional agriculture. These differences are due to the requirements of the building for the installation and operation of the i-RTG. In this sense, the i-RTG structure had to comply with safety regulation from the Spanish

Table 3.1 Summary of tools and methods applied in the project depending on the scale of analysis

Scale	Item		Tools
Building	Characterization of the system	Infrastructure	Direct measurements, architectural project, interview with architects (Sanyé-Mengual et al. 2015b)
		Food production	Measurement of the quantity of production and biomass generated and the concentration of nutrients using spectroscopy (ICP-OES)
		Water	Measurement of the quantity of water using a flow metre, the pH and the electric conductivity using field equipment and the concentration of nutrients using ion chromatography and spectroscopy (ICP-OES)
		Energy	Measurements using temperature and humidity sensors and pyranometers and application of the Building Energy Modelling Software (TAS) (Nadal et al. 2017)
		CO_2 emissions	Measurements using CO_2 sensors, anemometers and pyranometers
		Biomass	Application of the carbon elemental analysis, thermogravimetry, metal content analysis, development of material samples and characterization of material properties
	Environmental assessment		Life cycle assessment (LCA) (ISO 14040:2006; ISO 14044:2006)
	Economic assessment		Life cycle costing (LCC) (ISO 15686-5:2008)
Urban	Social perception analysis		Semi-structured interviews, surveys
	Urban integration		Application of Geographic Information Systems (GIS), TASI-600 and Leica ALS50-II airborne sensors, LWIR (Long Wave Infrared) and LIDAR (Light Detection and Ranging) data

Technical Code of Edification (CTE) [RD 314/2006 (BOE 2006)], ensuring the resistance to the wind and the safety of the users and the citizens the public space around the building. Therefore, the amount of steel in the skeleton of the greenhouse was greater, and the structure of the i-RTG requires relatively more resources compared to a multi-tunnel ground-based greenhouse in Almeria (Spain) (Montero et al. 2011). Furthermore, the covering materials of the greenhouse were adapted to the requirements of the fire safety standards [RD 2267/2004 (BOE 2004), Law 3/2010 (BOE 2010)]. The design of a conventional Mediterranean greenhouse includes a cover made of low-density polyethylene (LDPE), leading to a light and resource-efficient greenhouse design, but this material does not comply with fire safety laws, and polycarbonate had to be applied instead. Such change involved both positive and adverse effects. On the one hand, polycarbonate has a longer lifespan than LDPE, decreasing the maintenance requirements of the i-RTG. On the other hand, the use of resources, the environmental impacts and the economic costs of the cover increased, as polycarbonate is more expensive and thicker than LDPE (Sanyé-Mengual et al. 2015c).

3.5.2 Food Production System

The plant species selected for the cultivation in the i-RTG were tomato (*Solanum lycopersicum*) and beans (*Phaseolus vulgaris*). The crops were grown in perlite bags (i.e. inert base), and an open hydroponic system was used for its irrigation, supplying the necessary water and nutrients to plants through drippers. This system has the advantage of reducing the weight that the building structure has to support in comparison with soil. The irrigation was done by an automatic system with a nutrient solution injected through drippers with a flow of 2 L/h. For an adequate irrigation, the drainage of the crop was measured daily, and irrigation was adjusted accordingly, keeping the leachates (excess of irrigation) at 30–40%, which is common practice. Samples were periodically collected from the nutrient solution and the leachates. The concentration of nutrients in these samples was measured for chloride, nitrite, nitrate, phosphate, sulphate, calcium, potassium and magnesium. Moreover, the pH and EC were measured to obtain an immediate feedback of the total nutrients in the solution to adjust it if necessary.

Samples of the biomass and the fruits of the plant were collected for biochemical measurements when the plants were at full maturity. For the measurement of the quality parameters of tomato production a protocol was implemented between two and six times per crop, harvesting six representative tomatoes (average sized, normal shaped and non-perimeter) from three different lines. Various quality parameters were measured including the diameter, length, weight and dissolved sugar (Brix). For the bean crops, the variety selected has a single period of production and maturation, and thus only four samples of 500 g each were collected during the crops, determining the weight and length of each sample pod. These samples of production were dried at 60 °C, and their dry mass was weighed and delivered for the analysis of the nutrients contained. At the end of the crop cycle, the nutrients used by plants in stems, leaves and roots were calculated weighing the biomass of six plants randomly selected from tomato and bean crops and were also delivered for the analysis of the nutrients contained.

Results show that the cultivation of tomatoes in winter is possible, but the yield is lower than expected due to the lower solar radiation and temperature during this season. Moreover, low transmission values are found due to the additional structural elements required in the i-RTG in comparison with a conventional greenhouse made of glass or polyethylene film (Sect. 5.1). According to these results, the cultivation of beans was considered as an alternative during winter, as it is a less demanding solar crop due to its low height. The yield obtained during the crops was measured and classified as edible and nonedible production, as well as the waste biomass from pruning. More agronomic data regarding the crops conducted during the project can be found in Table 3.2 and pictures illustrating the crops are shown in Fig. 3.3.

Fig. 3.3 i-RTG with 2-week tomato crop (top left), i-RTG with a 2-month tomato crop (top right), i-RTG from the atrium of the building (bottom left), tomato production (bottom right)

3.5.3 Assessment of Resources (Nutrients) and Water Flows

Water is a key factor for resource efficiency in the i-RTG, which aims at reducing its consumption and the environmental impacts that result from it. As discussed above, the i-RTG holds a fertigation system that delivers water and nutrients to the crop, and thus the analysis of water flows is inextricably linked to the nutrient use efficiency. With the purpose of analysing these flows in the greenhouse, a balance of the water and nutrients used was conducted for each of the crops considered in the assessment.

Table 3.2 Information of the crops conducted in the i-RTG between 11/02/2015 and 14/09/2016

Crop (year, season)	Variety	Transplanting date	Duration (days)	Total yield (kg/m^2)
Tomato	®Arawak 2015	11/02/2015	164	16.2
	®Tomawak 2015	15/09/2015	167	3.4
	®Arawak 2016	08/03/2016	139	11.2
Beans	Low bush bean	14/09/2016	57	0.5

During the experimental phase of the study, the concentration of nutrients (i.e. nitrates, phosphates, sulphates, calcium, potassium, magnesium) was periodically measured in the nutrient solution supplied to the crop and in the leachates using ion chromatography. Samples of the nutrient solution were collected once per week and leachate samples were collected three times per week. The quantity of water supplied to the crop was also measured daily using a flow metre, and the volume of the leachates from the crop was estimated using two trays that collected the leachates from two perlite bags in the crop. Furthermore, the fruit production and crop biomass were weighted, and the concentration of the same nutrients considered for liquid samples was analysed for production, stems and leaves via spectroscopy (ICP-OES). The retention of nutrients in the substrate bags (in perlite) was measured at the end of every crop to quantify the nutrients retained. For each of the bags considered, the perlite was homogenised, and a sample was analysed using spectroscopy (ICP-OES) for sulphur, phosphorus, calcium, potassium, magnesium, boron, iron, zinc, copper, manganese and sodium.

The results showed that the balance is very different for each of the nutrients analysed. For instance, most of the nitrogen and potassium supplied (nearly 50%) was lost through the leachates, whereas the amount of phosphorus present in the leachates was relatively low, representing 25%. The rest of the nutrients were embedded in the production and biomass, except a certain percentage that was retained in the perlite that can reach 6% of the total supplied in the case of phosphorus.

At a wider level, the water flows in the ICTA-ICP building were analysed from a quantitative perspective based on the synergy with the i-RTG. The building stores rainwater harvested from a 900 m^2 roof in a 100 m^3 water tank that supplies crops and ornamental plants in the building. The rainwater collected and used in the greenhouse and the drinking water from the conventional supply network were measured using flow metres. The results from this quantification show that most of the water utilised in the i-RTG (between 80 and 90% of the total) is rainwater collected in the building, which allowed saving a high percentage of the drinking water demand of the building. Thus, the use of rainwater is key to enhance water self-sufficiency in UA.

3.5.4 Assessment of Energy Flows

The integration or symbiosis between the i-RTG and the building includes a thermal exchange between the two systems. This connection is able to reuse the residual heat and cold of the building and the greenhouse to improve the climatic comfort conditions of both systems. The benefits of the i-RTG concept must be characterised to analyse its overall thermal performance. With this purpose, the temperature and other climatic variables were monitored in the greenhouse and outdoor environments. The monitoring design consists of devices uniformly distributed in sections of the i-RTG and in other parts, which are placed in vertical supports and measure daily temperature and humidity.

The results show that indoor night temperatures in the i-RTG are 10 °C higher than outside in winter, 7.5 °C in spring and 5 °C in summer. This higher temperature is essential to ensure crop productivity, as winter nights reach negative temperatures in this geographic area, thereby increasing the risk of mortality of vegetables. This difference seems to be mainly a consequence of the thermal inertia of the building and its materials, especially due to the concrete slab floor. This thermal inertia increases the benefits of this type of agro-urban form during winter, as it eliminates the need for mechanical heating based on fossil fuels (Nadal et al. 2017).

The i-RTG had better average temperatures (16.5–25.9 °C) than a freestanding greenhouse in a Mediterranean area (Nadal et al. 2017). The average i-RTG temperatures were within FAOs recommended/optimum range for Mediterranean horticultural closed systems, from 14 to 26 °C. The effect of thermal inertia on the thermal behaviour of the i-RTG resulted in a useful source of heat during nights. This heat offers a thermal advantage when compared to conventional greenhouses, where indoor and outdoor temperatures are similar at night, and there is a need to heat the greenhouse actively. Hence, the thermal metabolism of the i-RTG provided ideal conditions for crop growth almost all year and offered a substantial advantage in winter when compared to freestanding greenhouses. Finally, the productivity of the crops (Sect. 5.2) was partly a reflection of the stable and ideal rank of the annual thermal conditions of the i-RTG.

3.5.5 Managing CO_2 Emissions

The concentration of CO_2 in buildings such as offices and households can rise to 2500 ppm (ACGIH 1991). Concentrations higher than 800 ppm in indoor environments may have a negative effect on human health (Seppanen et al. 1999), and thus high ventilation rates are usually required to reduce these concentrations. The residual air from the ventilation systems is generally released to the atmosphere. However, the ICTA-ICP building was designed to take advantage of this waste air injecting it into the i-RTG. Hence, the air with high CO_2 levels generated in the labs and offices of the building can be used for carbon enrichment in the i-RTG, which increases CO_2 concentration in greenhouses over 500 ppm to stimulate plant productivity (Mortensen 1987). The previous literature demonstrates that, for example, CO_2 enrichment applied in tomato crops can increase crop yields by 18 and 30%, depending on the tomato variety (Yelle et al. 1990).

A detailed study of the ventilation system at the ICTA-ICP building revealed that the air injected into the i-RTG came only from the labs on the third floor but not from the offices of the building. Since the occupation in the labs was lower than in the offices, the CO_2 concentration of the air injected in the i-RTG was low. Moreover, as the labs needed to be under overpressure conditions, high air renovation rates were required, which did not allow to accumulate CO_2. Therefore, the residual air had not enough CO_2 concentration to be used for the carbon enrichment. Nevertheless, it was detected that the CO_2 concentration in the offices was sufficiently high to be utilised for this purpose. It would be of interest to evaluate new processes that produce CO_2 and inject it into the i-RTG to boost vegetable production.

3.5.6 Assessment of the Biomass Production

As a consequence of the development of UA in cities (Zezza and Tasciotti 2010), the generation of waste biomass is foreseen to increase in the future. Conventional waste management scenarios for agricultural biomass are made up of the utilisation to feed farm animals, energy valorization (i.e. incineration) or compost production. Although these solutions are sustainable forms of waste management, they release the biogenic carbon captured by crops into the atmosphere.

New waste management practices are required to ensure the sequestration of the CO_2 captured by agricultural wastes. The creation of new products with these agricultural wastes is an alternative solution to the current waste management options. The potential to create new products using the biomass from tomato plants grown in UA was evaluated. Within a circular economy approach, this strategy could provide the following advantages:

- Avoid the management of UA wastes and its corresponding environmental impacts.
- Produce new products with local materials, which require fewer logistics and have a low carbon footprint.
- Store the carbon emissions captured by UA crops.
- Avoid the depletion of natural resources.

Two products made with biomass as a by-product from tomato plants in the i-RTG were assessed, i.e. biochar for soil amendment and insulation material for the construction sector. The functional units selected for the assessment were (1) the production of one tonne of biochar and (2) the mass (kg) of insulation material required to provide a thermal resistance with an R-value of 1 m^2 K W^{-1} for a surface of 1 m^2. The biogenic carbon, which refers to the carbon emissions that are captured by crops and remain in the by-products, was also included in the assessment. The calculation of the carbon footprint is necessary to determine if the carbon emissions fixed through these by-products compensate the emissions generated during their production. These calculations included the transportation of waste from the i-RTG to a production plant, the energy consumption, possible additional raw materials used and the transportation of the final product to the consumption point.

The results show that the carbon footprint of both products was negative, which means that the CO_2 stored within the by-products is higher than the emissions released for its production and transportation. However, the logistics had to be simple to guarantee this negative carbon footprint. In the case of biochar, further research was still required to guarantee the technical feasibility of producing such by-products with residual tomato plants. For the insulation material, more samples and tests should be implemented to reduce its density and to improve its thermal properties.

3.6 Environmental and Economic Studies

3.6.1 Environmental Impacts of i-RTG Systems

The environmental burdens of the i-RTG operation phase were assessed using the LCA methodology. With this purpose, three crops were conducted and monitored in the i-RTG lasting more than 15 months in total. The objective of the assessment was to calculate the environmental impacts of tomato production in the i-RTG and to detect the environmental hotspots of the system. Another goal of this analysis was to compare the impacts of the i-RTG production with those of a conventional greenhouse tomato production.

The functional unit considered in the assessment of the system as well as in the comparison with the conventional greenhouse was 1 kg of tomatoes. The whole life cycle was considered, including the materials, transport, construction, use and disassembly of the i-RTG. The comparison was conducted considering the consumption of 1 kg of tomato in Barcelona being produced in the i-RTG and 1 kg being produced in a standard greenhouse in Almeria (800 km distance), where more than 50% of the tomatoes eaten in Barcelona are grown (MercaBarna 2017). The loss of product in the transport (around 10% in weight due to refrigeration) and the packaging used were accounted for in the comparison. In the case of the i-RTG, no transport, loss of product or packaging were considered, provided that users from the building take production at will using their bags and carry it home or eat it on site.

Some key factors were identified for the design of future i-RTG projects. The main drawback of the i-RTG is the shadows from elements of the building structure, which reduce the production and in turn affect the environmental impacts. This should be an imperative factor to consider for the selection of the i-RTG location. The use of fertilisers in the fertigation has shown being the main impacting element of the system, contributing to more than 25% of the impact in four out of six impact indicators. The installation of a closed hydroponic system with recirculation of nutrients might be of interest to reduce the fertilisers used and the water demand. However, this irrigation system would require more mechanisms for pumping, storing and disinfecting the leachates, becoming more complex and challenging to manage.

Furthermore, the environmental impacts of the greenhouse structure and the rainwater harvesting system of the building (which are part of the infrastructure of the i-RTG) contribute with more than 30% to the total environmental impacts of the system. The auxiliary equipment for the irrigation did not show significant environmental impacts, representing less than 5% of the impacts in six out of seven impact categories. Both the greenhouse structure and the rainwater harvesting system have been exaggerated in size, and the material used for its manufacture could have been lower if a study for its optimisation would have been conducted during the design phase. For instance, the 100 m^3 rainwater storage tank could have been replaced by a 20 m^3 tank which would have collected 90% as much rainwater as the prior one, reducing its environmental impacts substantially. The reason for these results is that the limiting factor in rainwater harvesting systems is usually the catchment surface, not the tank size.

Production from the i-RTG has shown having between 50 and 75% lower environmental impacts than conventional greenhouses in six out of the seven impact categories analysed, mainly due to the lower needs for packaging and transport. This fact proves that, in addition to increasing food security, i-RTGs have the potential to reduce the environmental impacts of food production.

3.6.2 *Economic Costs of i-RTGs*

The operation phase of the i-RTG was assessed from an economic perspective using the life cycle costing (LCC) method and focusing on the first of the crops conducted in the i-RTG, which lasted 6 months. This economic assessment aims to evaluate the cost breakdown of the operation stage to identify the main elements contributing to the final production costs and to discuss the comparison between the economic performance of the i-RTG and other production systems.

As for the environmental assessment, 1 kg of tomato production was selected as the functional unit for the calculation of the final costs and for further comparison. To emulate the real cost of the crop, the costs of the monitoring of the system were not accounted. Moreover, the full price of the elements bought was considered, compensating the tax-free products supplied to the university.

In total, the final cost of production in the i-RTG was 10.5 €/kg of tomato, which is more than double the current cost of beef tomato in the market. Labour accounts for 86% of the total and, therefore, has the highest contribution to this final cost. This is mainly because the staff in charge of the agricultural tasks had no previous knowledge of greenhouse management and its efficiency was much lower than in conventional greenhouses. Moreover, the salary of the staff in the university is much higher than the wages of workers in Southern Spain greenhouses.

However, direct comparison of the i-RTG costs and conventional production in standard greenhouses is complicated. Conventional greenhouses benefit from economies of scale, and their main objective is economic profit; in contrast, the i-RTG is a pilot system at a small scale whose main purpose is to conduct research. The application of the i-RTG technology at an industrial scale would contribute to reduce the economic costs of production and make this technology economically attractive. Moreover, the application of this innovative system in the future might not pursue economic benefit but environmental and social positive impacts.

3.7 Social Studies on the i-RTG Implementation at the Urban Scale

The potential implementation of rooftop agriculture can be constrained by public acceptance and opinion, particularly in the case of business-oriented initiatives that seek a market niche and specific consumers (Sanyé-Mengual et al. 2016; Specht et al. 2016). A social analysis was performed to assess the perception of the different stakeholders of the area of Barcelona regarding the potential implementation of rooftop greenhouses. A total of 25 stakeholders were interviewed, including various offices of the local and regional administration, architects and engineers, peri-urban farming actors, urban gardeners and practitioners, NGOs, and food co-ops. Semi-structured interviews were used to explore the discourses behind the implementation of UA and the potential integration of rooftop agriculture in the development and planning of Barcelona.

The results unveiled a strong relationship between the perception and acceptance of rooftop farming and the preconceived definitions and concepts of UA (Sanyé-Mengual et al. 2014a). Some stakeholders highlighted the social role of UA in cities like Barcelona, undervaluing at the same time the potential contribution of UA to urban food security. Thus, these groups of stakeholders did not accept the use of high-tech and resources for the development of UA, outlining the need to be in contact with the soil instead of conquering the roofs. On the contrary, some stakeholders valued the potential food production of rooftop agriculture and even underpinned the business opportunity for new entrepreneurs. In general, the stakeholders identified several environmental, social and economic benefits of rooftop farming and rooftop greenhouses, which can strongly contribute to the short-term urban sustainability. However, various risks and barriers were tied to rooftop farming, such as the possible gentrification effect or the food safety uncertainty due to air and soil contamination (Sanyé-Mengual et al. 2016).

At the same time, the social perception, the marketing preferences and the willingness to pay were evaluated by potential consumers of products grown in the i-RTG, focusing on tomatoes. With this purpose, 227 surveys were carried out by supplying tomato samples. The results indicated that 86% of the respondents considered this tomato a consumption option. Marketing in supermarkets was the dominant pattern of purchase, and the willingness to pay was below the market price. It was also perceived that it is necessary to inform the population about the processes and benefits of this type of UA products so that consumers are aware of the environmental savings that their acquisition and consumption implies.

3.8 Towards the Integration of i-RTGs in Urban Planning

The potential area for the implementation of UA in the urban planning and its production is one of the main research topics under assessment. This issue has been covered by several authors analysing the potential for self-sufficiency of urban areas, particularly for cities in the United States (Ackerman 2011; Grewal and Grewal 2012; McClintock et al. 2013). In this sense, the Fertilecity project aimed to account for the potential implementation of i-RTGs in Barcelona.

Geographic Information Systems (GIS) were used in combination with LCA to evaluate the potential area, the social benefits and the environmental impacts (Sanyé-Mengual et al. 2015a). The resulting GIS-LCA guide focused on the implementation of the i-RTGs in industrial and logistic parks where the i-RTGs could take advantage of the residual flows of industrial activities, as well as from the supply-chain of logistics services. The industrial park of Zona Franca, located at the South of Barcelona city (Spain) is this analysis case of study due to its geographical position and its great extension.

Experts on agronomy, architecture, engineering and environmental sciences defined a set of prerequisites for implementing the i-RTGs on the roof of industrial buildings (including technical, economic, legal and agricultural criteria). By employing GIS, the roofs of the Zona Franca were characterised according to these features to determine the technically and economically feasible roofs for a short-term i-RTG implementation. Eight percent of the roofs of the industrial park were assigned a short-term potential. The assessment of the potential tomato production and the avoided environmental burdens of substituting the supply chain of conventional tomatoes imported from Almeria (Spain) indicated that almost 2000 tonnes of tomatoes could be produced, supplying the demand of 150,000 inhabitants of Barcelona. This production could avoid 65.25 t of CO_2 eq./m^2 of the rooftop greenhouse (Sanyé-Mengual et al. 2015a).

The application of the i-RTG technology was also analysed for developing countries. The guide was adapted by considering a social approach and the potential implementation of rooftop greenhouses in social neighbourhoods in Quito (Ecuador), was evaluated. The results showed that 33% (7.70 ha) of the neighbourhood rooftops had a short-term feasibility to become rooftop greenhouses, with the potential to produce 1579 and 56 tonnes/year of tomato and lettuce, respectively.

Complementarily, the possibility of automating the data collection using airborne sensors for the acquisition of information regarding the rooftops is also analysed. The TASI-600 and Leica ALS50-II sensors and LWIR (Long Wave Infrared) and LIDAR (Light Detection and Ranging) sensors enabled the acquisition of quantitative data regarding the main physical characteristics (e.g. dimensions, slope, material, solar radiation) of the roofs. The main advantages of this technology—in contrast to more traditional methods—are that information can be collected in real time, with high precision and digitally.

3.9 Conclusions

In the field of sustainability, multidisciplinary approaches are key to ensuring the complete understanding of problems and solutions towards more resilient cities. The Fertilecity project provided a unique vision of the implementation of UA in cities, specifically RTGs and i-RTGs. The first results of this research are summarised in this chapter, which highlights the benefits, constraints and future pathways in the application of UA at the building and urban scale as a climate change mitigation and food security strategy. In general, the residual energy, rainwater and CO_2 exchanges between the building and the i-RTG proved to be positive regarding heat comfort, resource availability and carbon fixation. This strategy might also enable the creation of new market niches, such as the commercialization of RTG-grown produce or by-products. This is still a field to explore, but it might add up to the inherent value of i-RTG strategies. Social and legal issues were known constraints to the implementation of UA at the city scale. For this reason, there is a need for solutions that negotiate these limitations and enhance the production of food in urban areas.

These results and methods are still novel, especially in the Mediterranean context. For this reason, it is not straightforward to obtain an exact equivalent system that performs the same function. Land use impacts, additional building-greenhouse synergies and user performance might have an effect on the results. Hence, this study is only the first step towards a complete detailed set of research analyses that shed light on the sustainability benefits and shortcomings of integrating UA into buildings.

Acknowledgements The authors thank the Spanish Ministry of Economy and Competitiveness for awarding and funding the Fertilecity project (CTM2013-47067-C2-1-R) and (CTM2016-75772-C3-1-2-3-R) (MINECO/FEDER,UE). The authors are also grateful for the research fellowships awarded to D. Sanjuan-Delmás, P. Llorach-Massana, and M. Ercilla-Montserrat by Agaur—Generalitat de Catalunya (FI-DGR 2014, 2015, 2016); A. Petit-Boix and E. Sanyé-Mengual, by the Spanish Ministry of Education (FPU13/01273; AP2010-4044), and A. Nadal, by the National Council for Science and Technology of Mexico (CONACYT) and the Council for Science, Innovation and Technology, State of Yucatan (CONCIYTEY). Thanks to the Catalan Government for the SGR funds (2014 SGR 1412). We acknowledge the financial support from the Spanish Ministry of Economy and Competitiveness, through the 'María de Maeztu' programme for Units of Excellence in R&D (MDM-2015-0552).

References

ACGIH (1991) Documentation of the threshold limit values and biological exposure indices, 6th edn. Acgih Publications

Ackerman K (2011) The potential for urban agriculture in New York City: growing capacity, food security, and green infraestructure. Urban Design Lab, Earth Institute, Columbia University, New York

Ajuntament de Barcelona (2016) Statistical yearbook of Barcelona city. Year 2016 > Demography and population. http://www.bcn.cat/estadistica/angles/dades/anuari/cap02/C020104.htm

Architects HA (2011) Final Project documents of ICTA http://www.harquitectes.com/projectes/centre-recerca-uab-icta-icp/

Arredondo D (2013) Agricultura en la Ciudad: de la Utopía a la Conciencia de Lugar, Granada, Universidad de Granada

BOE (2004) REAL DECRETO 2267/2004, de 3 de diciembre, por el que se aprueba el Reglamento de seguridad contra incendios en los establecimientos industriales. Boletín Of del Estado 303:41194–41255

BOE (2006) REAL DECRETO 314/2006, de 17 de marzo, por el que se aprueba el Código Técnico de la Edificación. Boletín Of del Estado 74:11816–11831

BOE (2010) Ley 3/2010, de 18 de febrero, de prevención y seguridad en materia de incendios en establecimientos, actividades, infraestructuras y edificios. Boletín Of del Estado 89:32918–32943

Cerón-Palma I, Sanyé-Mengual E, Oliver-Solà J et al (2012) Barriers and Opportunities Regarding the Implementation of Rooftop Eco. Greenhouses (RTEG) in Mediterranean Cities of Europe. J Urban Technol 19:87–103. https://doi.org/10.1080/10630732.2012.717685

Chalana M (2014) Chandigarh: City and periphery. J Plan Hist 14:62–84

Contreras E, Castillo I (2015) Guia de terrats vius i cobertes verdes. Ajuntament de Barcelona. http://hdl.handle.net/11703/86542
da Cunha JMP, Rodríguez-Vignoli J (2009) Crecimiento urbano y movilidad en América Latina. Rev Latinoam Población 3:27–64
Deely SC, Dodman DC, Hardoy JC, Johnson CC (2010) World disasters report 2010: Focus on urban risk
dos Santos MJPL (2016) Smart cities and urban areas—Aquaponics as innovative urban agriculture. Urban For Urban Green 20:402–406. https://doi.org/10.1016/j.ufug.2016.10.004
European Commission (2010) Making our cities attractive and sustainable—How the EU contributes to improving the urban environment. Publications Office of the European Union, Luxemburg
Fraunhofer UMSICHT (2011) Harvesting on urban rooftops—demo center at Fraunhofer inHaus. Fraunhofer institute for Environmental, Safety and energy technology UMSICHT, Oberhausen
Grewal SS, Grewal PS (2012) Can cities become self-reliant in food? Cities 29:1–11
ISO 14040:2006 Environmental management—life cycle assessment—principles and framework
ISO 14044:2006 Environmental management—life cycle assessment—requirements and guidelines
ISO 15686-5:2008 Buildings and constructed assets—sevice-life planning—part 5: life-cycle costing
Janick J (2002) Ancient Egyptian agriculture and the origins of horticulture. In: Acta horticulturae, pp 23–39
Joffe H, Smith N (2016) City dweller aspirations for cities of the future: how do environmental and personal wellbeing feature? Cities 59:102–112. https://doi.org/10.1016/j.cities.2016.06.006
Lakkireddy KKR (2012) Role of hydroponics and aeroponics in soilless culture in commercial food production. Res Rev J Agric Sci Technol 1:26–35
Lufa Farm (2017) Our vision is a city of rooftop farms. https://lufa.com/
Lugaric L, Krajcar S (2016) Transforming cities towards sustainable low-carbon energy systems using emergy synthesis for support in decision making. Energy Policy 98:471–482. https://doi.org/10.1016/j.enpol.2016.09.028
McClintock N, Cooper J, Khandeshi S (2013) Assessing the potential contribution of vacant land to urban vegetable production and consumption in Oakland, California. Landsc Urban Plan 111:46–58. https://doi.org/10.1016/j.landurbplan.2012.12.009
McKay G (2011) Radical gardening: politics, idealism and rebellion in the garden. Frances Lincoln, London
MercaBarna (2017) Database of products' trade: fruits & vegetables—year 2016 [online]. http://www.mercabarna.es/serveis/estadistiques-productes/. Accessed 20 Feb 2017
Mitchell RG, Spliethoff HM, Ribaudo LN et al (2014) Lead (Pb) and other metals in New York City community garden soils: factors influencing contaminant distributions. Environ Pollut 187:162–169. https://doi.org/10.1016/j.envpol.2014.01.007
Montero J, Antón A, Torrellas M et al (2011) EUPHOROS Deliverable 5. Report on environmental and economic profile of present greenhouse production systems in Europe. European Commission FP7 RDT Project Euphoros (Reducing the need for external inputs in high value protected horticultural and ornament)
Morán A, Aja A (2011) Historia de los huertos urbanos. I Congreso Estatal de Agricultura Ecológica Urbana y Periurbana
Mortensen LM (1987) Review: CO_2 enrichment in greenhouses. Crop responses. Sci Hortic (Amsterdam) 33:1–25. https://doi.org/10.1016/0304-4238(87)90028-8
Nadal Fuentes A (2015) Agricultura urbana en el marco de un urbanismo sostenible. Temes de disseny 31:92–103
Nadal A, Llorach-Massana P, Cuerva E et al (2017) Building-integrated rooftop greenhouses: an energy and environmental assessment in the mediterranean context. Appl Energy 187:338–351. https://doi.org/10.1016/j.apenergy.2016.11.051
Pons O, Nadal A, Sanyé-Mengual E et al (2015) Roofs of the future: rooftop greenhouses to improve buildings metabolism. Procedia Eng 123:441–448. https://doi.org/10.1016/j.proeng.2015.10.084

Sanyé-Mengual E, Anguelovski I, Oliver-Solà J et al (2014a) When the perception and development of urban rooftop farming depend on how urban agriculture is defined: examining diverging stakeholders' experiences and views in Barcelona, Spain. In: Roggema R, Keefe G (eds) "Finding spaces for productive spaces" 6th AESOP sustainable food planning conference. VHL University of Applied Sciences, Velp, pp 491–504

Sanyé-Mengual E, Llorach-Massana P, Sanjuan-Delmás D et al (2014b) The ICTA-ICP rooftop greenhouse lab (RTG-Lab): closing metabolic flows (energy, water, CO_2) through integrated rooftop greenhouses. In: "Finding spaces for productive cities," 6th AESOP sustainable food planning conference, pp 692–701

Sanyé-Mengual E, Cerón-Palma I, Oliver-Solà J et al (2015a) Integrating horticulture into cities: a guide for assessing the implementation potential of rooftop greenhouses (RTGs) in industrial and logistics parks. J Urban Technol 22:87–111

Sanyé-Mengual E, Oliver-Solà J, Montero JI, Rieradevall J (2015b) An environmental and economic life cycle assessment of rooftop greenhouse (RTG) implementation in Barcelona, Spain. Assessing new forms of urban agriculture from the greenhouse structure to the final product level. Int J Life Cycle Assess 20:350–366. https://doi.org/10.1007/s11367-014-0836-9

Sanyé-Mengual E, Anguelovski I, Oliver-Solà J et al (2016) Resolving differing stakeholder perceptions of urban rooftop farming in Mediterranean cities: promoting food production as a driver for innovative forms of urban agriculture. Agric Human Values 33:101–120. https://doi.org/10.1007/s10460-015-9594-y

Säumel I, Kotsyuk I, Hölscher M et al (2012) How healthy is urban horticulture in high traffic areas? Trace metal concentrations in vegetable crops from plantings within inner city neighbourhoods in Berlin, Germany. Environ Pollut 124–132

Seppanen OA, Fisk WJ, Mendell MJ (1999) Association of ventilation rates and CO_2 concentrations with health and other responses in commercial and institutional buildings. Indoor Air 9:226–252. https://doi.org/10.1111/j.1600-0668.1999.00003.x

Specht K, Siebert R, Thomaier S (2016) Perception and acceptance of agricultural production in and on urban buildings (ZFarming): a qualitative study from Berlin, Germany. Agric Hum Values 33:753–769. https://doi.org/10.1007/s10460-015-9658-z

Spudic S (2007) The new victory garden. Royal Horticultural Society Dissertation, Wisley, Diploma in Practical Horticulture

The Vinegar Factory (2017) Eli's Vinegar factory. http://www.elizabar.com/-C24.aspx

United Nations (2014) World's population increasingly urban with more than half living in urban areas. Dep Econ Soc Aff 2014–2016

Yelle S, Beeson RC, Trudel MJ, Gosselin A (1990) Duration of CO_2 enrichment influences growth, yield, and gas exchange of two tomato species. J Am Soc Hortic Sci 115:52–57

Zaar M-H (2011) Agricultura urbana: algunas reflexiones sobre su origen e importancia actual

Zezza A, Tasciotti L (2010) Urban agriculture, poverty, and food security: empirical evidence from a sample of developing countries. Food Policy 35:265–273. https://doi.org/10.1016/j.foodpol.2010.04.007

Chapter 4
Urban Home Gardens: A Sustainable Conservation Model for Local Plants Based on Mexican Urban Agri-Horticultural Practices

W. S. Poot-Pool, W. Cetzal-Ix, S. K. Basu, E. Noguera-Savelli and D. G. Noh-Contreras

Abstract Home gardens represent an important agri-forestry systems widely practiced across the Yucatan Peninsula of Mexico successfully combining tree and shrubs; along with annual and perennial agricultural crop species and farm animals. They are found around houses in the rural settlements of Mexico mostly under the family management. However, the practice has been successfully extended into the urban settlements of Mexico including those in the Yucatan Peninsula as a highly popular environment-friendly, sustainable agri-horticultural practice combining indigenous plants suitable for food, fodder, fertilizer, forage and medicinal purposes. The urban home or family garden contributes significantly towards local food security and nutrition, strengthening family economy and in the conservation of regional and local agri-biodiversity; and could successfully serve as a sustainable, green, environmental model. Recently, the local Mexican administration has been promoting for developing school home gardens as strategies to raise awareness regarding conservation of regional flora and food security among school children further strengthening the concept of urban home gardens in Mexico. In this chapter, we highlight the history, development, structure, nature, species used and their uses along with importance of home gardens in Mexico with special emphasis to Yucatan Peninsula highlighting some specific peri-urban and urban case studies.

Keywords Home gardens · Family gardens · Agri-forestry · Agri-biodiversity Agri-horticultural · Conservation · Yucatan peninsula · Mexico

W. S. Poot-Pool · W. Cetzal-Ix (✉) · D. G. Noh-Contreras
Instituto Tecnológico de Chiná, Calle 11 Entre 22 Y 28, Colonia Centro Chiná, 24050 Campeche, Mexico
e-mail: rolito22@hotmail.com

S. K. Basu
Department of Biological Sciences, University of Lethbridge, Alberta, CanadaT1K 3M4

E. Noguera-Savelli
Catedrática CONACYT, Colegio de Postgraduados Campus Campeche, Carretera Haltunchén-Edzná Km. 17.5, Sichochac, Champotón, Campeche, Mexico

D. Nandwani (ed.), *Urban Horticulture*, Sustainable Development and Biodiversity 18, https://doi.org/10.1007/978-3-319-67017-1_4

4.1 Introduction

Urbanization refers to the movement of people from rural areas to the cities (Pickett et al. 2001). In peri-urban and urban areas, the main activities of people are no longer agriculture; but various services sectors and the local industries (Molebatsi et al. 2010). In urban areas, the population density is high; and therefore the availability of resources is quite limited, causing enhanced pressure on the scanty available natural resources (Pauchard et al. 2006). There is significant reduction in species richness due to species loss, habitat fragmentation, and disturbance and/or destruction of natural habitats. Although there is an increase of species in family gardens or home gardens in the form of predominantly nonnative plants (flora) that apparently diminishes the local and regional biodiversity (Goddard et al. 2009). According to Williams et al. (2009) the effects of urbanization on vegetation is summarized in four aspects: habitat transformation, habitat fragmentation, environmental alteration, and human preferences. In the literature (Nemudzudzanyi et al. 2010; Molebatsi et al. 2010; Poot-Pool et al. 2015) it has been mentioned that urbanization can affect family home gardens; the floristic composition and tree structure of a family home gardens in rural and semi-rural areas is not the same as in peri-urban and urban areas.

Home gardens are agri-forestry systems widely practiced in the Yucatan Peninsula of Mexico (Fig. 4.1) that combines tree species and shrubs with annual and perennial agricultural crops or animals. They are found around houses and under family management (Torquebiau 1992). The production of family home gardens in rural localities is oriented to subsistence and generates a wide diversity of fruit and medicinal trees to meet the basic needs of families, as well as contributes to the conservation of agri-biodiversity (Fernández and Naír 1986; Soemarwoto and Conway 1992; Lamont et al. 1999). However, the practice of home gardens can also be successfully applied in peri-urban and urban areas as a highly popular environment-friendly, sustainable agri-horticultural practice combining indigenous plants suitable for food, fodder, fertilizer, forage, and medicinal purposes (Fig. 4.1).

A study of peri-urban home gardens by Nemudzudzanyi et al. (2010) from Africa showed that people value biodiversity as a component of culture and livelihoods. However, a large number of the exotic species reported were predominantly fruit species that are edible, medicinal and used as natural shade. In another study from Africa (Molebatsi et al. 2010), it was clearly shown that in peri-urban home gardens the percentage of species with ornamental use is higher than that in rural home gardens. From the perspective of local biodiversity conservation, it has been mentioned that peri-urban home gardens conserve biodiversity less than that in the rural areas (Molebatsi et al. 2010). It has been mentioned in the literature that home gardens near urban areas are less dependent on their production; which means that there are fewer plants with different uses in the home gardens (Abdoellah et al. 2002; Peyre et al. 2006).

This chapter focuses on the history, development, structure, nature, species used and their uses along with importance of home gardens in Mexico with special emphasis to Yucatan Peninsula highlighting some specific peri-urban and urban case studies.

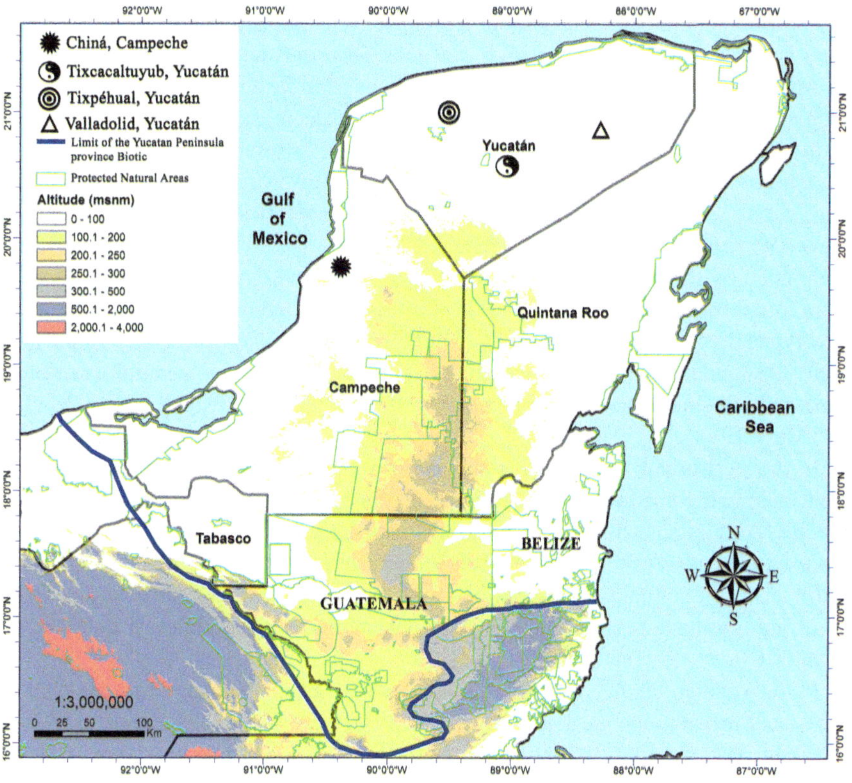

Fig. 4.1 Urban home gardens in the Yucatan Peninsula, Mexico. A-E. Ornamental plants (asparagus, ferns, orchids, palms plants). E. Medicinal and aromatic plants. F. Edible plants. *Photo credit* W. S. Poot-Pool

4.2 Size of Peri-urban Versus Rural Home Gardens

The size of the urban home gardens is smaller in comparison to that of the rural home gardens. However, it is not a reason for the species richness to be negatively affected; since strategies can be applied to maintain a diversity of species in small areas, as well as a great number of nonnative species in the peri-urban home gardens (Cilliers et al. 2007). The function of home gardens may vary depending on their locations (whether they are found in rural, peri-urban or urban) indicating that they can vary with respect to their floristic composition and structure.

For example, Abebe et al. (2010) conducted a study on 144 home gardens in Ethiopia and found two important trends. On one hand, the trend in home gardens was related to commercial crops; as a result of commercial networks that offer better income generation due to the proximity of local markets and steady supply of local customers. On the other hand, an increase in the demand for annual food crops (like corn and potatoes) changed the orientation of home garden production system, resulting in the gradual decrease of local species diversity and a less stratified canopy.

Another example is the study of home gardens from Java, Indonesia; where factors such as population growth and access to a market decreased the number of species and diversity in general (Michon and Mary 1994). Several types of home gardens were reported; some were dominated by monocultures, oriented to commercialization, such as coffee (*Coffea arabica* L.), walnut (*Myristica fragrans* Houtt.) and clove (*Syzygium aromaticum* (L.) Merr. & L. M.Perry). While the remaining were dominated by the fruit and timber trees; and finally in some home gardens orchids were found for commercialization. The shift from tall and multilayer orchards to the new types involved the removal of several trees, (particularly the high stratum) and their subsequent replacement by cash crops. As a consequence, the number of home gardens with complex structural characteristics has decreased significantly (Michon and Mary 1994).

In the peri-urban areas with good access to markets; family or home gardens contribute to income generation through the sale of fruits, vegetables and ornamental species (Abdoellah et al. 2002; Soemarwoto and Conway 1992). Ornamental species represent a symbol of the owner's socioeconomic status since the home gardens also fulfill an aesthetic function (Kehlenbeck 2007). Hence, ornamentals produced in the home gardens have been found to have a consistent local market demand (Poot-Pool et al. 2015).

In the Yucatan Peninsula, only four studies have been conducted till date on peri-urban and urban home gardens (Fig. 4.2a-g). For example, in Tixpéhual and Tixcacaltuyub, Yucatan, it was observed that in the home gardens near the urban areas; the composition was ornamental and commercial species (Rico-Gray et al. 1990). In the Chiná town, Campeche, it was observed that the home gardens were dominated by ornamental species especially herbaceous (Poot-Pool et al. 2015). One might think that urban home gardens do not contribute to the conservation of agri-biodiversity. However, in home gardens in Valladolid municipality, Yucatán; Novelo-Chan (2007) observed that they have a high biodiversity. The author explained that it is probable that the movement of people increases the number of species and successfully add to the local species richness as well as the local and regional biodiversity.

4.3 Importance of Plants Present in Home Gardens in Peri-urban and Urban Areas

Plants play an important role in sustaining urban spaces (Colding 2007). They provide physical services such as temperature and flood control, carbon removal in the atmosphere; and social services such as adding to the aesthetic value as well as promoting general well-being by establishing a healthy, local ecosystem and an enriched natural environment (Lubbe 2011). Furthermore, plants also have a positive impacts on the quality of life, human, and domestic animal health (Goddard

Fig. 4.2 Peri-urban home gardens in the Yucatan Peninsula, Mexico. A-F. Ornamental plants. E. Edible, medicinal and aromatic plants *Photo credit* W. S. Poot-Pool

et al. 2009). Green spaces provide opportunities for people to interact with nature and also promote a wider interest in nature and natural conservation. Green spaces are becoming an important refuge for native biodiversity (local flora and fauna);

therefore, home gardens also have the potential within a city to contribute towards successful biodiversity conservation (Goddard et al. 2009).

Although individual home garden represents only a small space; however, the total contribution of home gardens in the city is significant when all these individual spaces are summated together (Gaston et al. 2005). Recent studies indicate that urban areas play an important role in the maintenance of biodiversity through the provision of living spaces and connecting networks between large spaces and other natural areas that can easily serve as wildlife corridors (Kehinde-Akinnifesi et al. 2010). Orchards can play an important role as an alternative space; that serves as a network connecting rural and urban spaces, thereby helping in broadening the opportunity of species richness and significantly establishing local biodiversity (Heckler 2004; Winklerprins 2002).

It has now been clearly established through several comprehensive studies that family or home gardens have the potential to contribute towards the conservation of biodiversity in both rural and urban areas in developed as well as developing and under-developed nations (Galluzzi et al. 2010; Watson and Eyzaguirre 2001). For example, in Niger, Africa a high diversity of species was found in large peri-urban home gardens (Bernholt et al. 2009). In another study on urban home gardens at São Luís, North Brazil, reported that urban areas serve as a bank of indigenous species including those that are at risk of extinction (Akinnifesi et al. 2010).

4.4 Model of Urban Home Gardens in Mexico

Recently, the local administration has been promoting for developing school home gardens as strategies to raise awareness regarding the conservation of regional flora and food security among school children further strengthening the concept of urban home gardens in Mexico. Given the current reality of environmental deterioration and its influence on health and food related issues; the Mexican government policies are now sincerely looking for solutions with the establishment of school home gardens (FAO 2010).

4.5 School Home Gardens in Mexico

School home gardens or school gardens are areas cultivated by pupils within or near the schools. Initially they were designed for scientific education and/or agricultural training or for the purpose of income generation for the local schools. However, given the demands for food to ensure good nutrition among students and several environmental issues in many areas; the objectives of school home gardens have been expanded to promote better nutrition, improved livelihood and in building awareness regarding the local ecosystem and environment (FAO 2010). In this context, the urgent need of various developing and under-developed countries for promoting better education, awareness and establishing quality ecosystem and environment is leading to new approaches worldwide on the possibilities and scope of school home gardens.

FAO (2010) outlines the needs and approaches to the global outlook of school home gardens with respect to better nutrition and food security, securing livelihood as well as ecological restoration and environmental protection. Global Warming and Climate Change poses a serious threat to global food security, resulting in high demand for food with alarming rise in food expenses for government around the planet. In this perspective, the conservation and sustainable use of available natural resources to guarantee environmental protection as well as food security; it is important to secure livelihoods for the most vulnerable and marginalized communities with greatest nutritional needs with more than one-third of infant mortality is attributable to under nutrition (FAO 2010). On the other hand, it is important to generate awareness regarding unhealthy eating habits that produce obesity at very young ages, leading to degenerative diseases since childhood.

According to the needs previously mentioned by the FAO (2010), it proposes the following approaches: solutions related to nutrition; models of family home gardens and school lunches. The school home gardens should be considered holistically as a part of the "School Ecology Project" that provides training, education and experience in food security and nutrition, sustainable agricultural practices, organic farming and horticulture. Expected benefits of home gardening and management in a school environment provides the child (student) an opportunity to learn how to grow different types of food so that they can know how to value it and include it in their family home gardens. It teaches them the importance of healthy diet and to be able to appreciate and directly experience the close relationship between better health and nutrition in conjunction with the protection of local biodiversity, ecosystems and natural environments. The direct contact with horticulture inculcates in the student the value of the consuming fresh fruits and vegetables cultivated by them; and it teaches them the nutritional value of quality food and to discard consuming junk food. The long-term goal of school gardens has been building awareness regarding food security and promoting better quality of life for the school children.

In the 1940s, the use of the school plot in rural schools was formalized in Mexico with the aim of promoting education for productive work, promoting cooperation between communities, between students and teachers, as well as the generation of resources for the continuity of educational work and support to teachers (CICEANA 2017). With the aim of educating school children regarding the importance of cultivating grains and vegetables through the use of sustainable agriculture techniques and to learn to care for the local environment; a dynamic project has been established in Mexico, entitled "My school plots produces" under the direction of the Ministry of Agriculture, Livestock, Rural Development, Fisheries and Food (SAGARPA, by its acronym in Spanish) and the National Union of Workers of Education (SNTE, by its acronym in Spanish). The project is developed with an intention to influence the education of more than 21,000 students, beginning in six states: Aguascalientes, Chihuahua, Baja California, Sinaloa, Tlaxcala, and Yucatán, and later extending it to all the country's entities with the objective of promoting knowledge and development of nopal, tomato, maize, barley, lettuce (SAGARPA 2017). On the other hand, it aims to rekindle the interest of young people in agricultural activities, seeking to reduce the age range of people who are currently in the country.

In states like Yucatan, school gardens had previously been established in 2013 as a municipal Integral Family Development (DIF, by its acronym in Spanish) program in the city of Mérida, with an intended objective of catering to approximately 7000 primary school children. In conjunction with programs such as backyard gardens set up in 2016, the goal has been to ensure the food security of the most vulnerable populations, to change eating habits by focusing on the production and consumption of vegetables and fruits, and the promotion of better environmental care for protecting vulnerable local ecosystems and regional and local biodiversity.

4.6 Case Study of a Comparison of Peri-urban Versus Rural Home Gardens in Campeche, Mexico

Poot-Pool et al. (2015) compared rural, semi-rural and peri-urban home gardens in Campeche, Mexico (Figs. 4.1 and 4.3a-g). The authors found that richness in trees and shrubs was 79 species in rural, 73 species in semi-rural and 53 species in peri-urban home gardens. As for herbaceous habit, 161 species were found in peri-urban, 104 species in semi-rural areas and 82 species in the rural home gardens. The composition of tree and shrub species among home gardens presented differences, although these share 38 species in common (Table 4.1); Rural and urban home gardens share 43 species, rural and semi-rural 58 species, urban and semi-rural 41 species (Table 4.1). The main species shared among the orchards are edible species of the family Annonaceae, Anacardiaceae, Rutaceae (Fig. 4.4a-l). The common ornamental species widely cultivated in the Yucatan Peninsula include Apocynaceae, Bignoniaceae, and Fabaceae (Fig. 4.4a-l).

As for the origin of the tree species, it had a higher percentage of native species in rural than in peri-urban home gardens; however, with respect to introduced or nonnative species, it was greater in the latter. With regard to herbaceous plants, introduced species in peri-urban were higher than that in the rural home gardens. As for the uses of tree species, the average percentage of fruit species was higher in peri-urban than in rural home gardens. The average proportion of ornamental herbaceous species was higher in peri-urban than in rural home gardens. In rural home gardens the number of forage, timber, wood, and other uses is higher than in peri-urban home gardens (Poot-Pool et al. 2015).

This study of rural and peri-urban home gardens, concluded that the composition and life strategies of a rural with respect to a peri-urban home gardens are not the same. Each family adapts their home garden according to their particular economic strategies and needs, which results in changes in the functions of family gardens and the family economy and in the structure and floristic composition of the home gardens. If a family works primary activities (such as agriculture); this will reflect

Table 4.1 Comparison of floristic composition or tree and shrub in rural, peri-urban and semi-rural home gardens in Campeche, Yucatan Peninsula, Mexico

Family	Scientific name	R	Pu	Sr	Uses	OD
Acanthaceae	*Tetramerium nervosum* Nees	–	–	1	2	N
Anacardiaceae	*Anacardium occidentale* L.			1	1	I
Anacardiaceae	*Mangifera indica* L.	1	1	1	1	I
Anacardiaceae	*Spondias purpurea* L.	1	1	1	1	I
Annonaceae	*Annona muricata* L.	1	1	1	1	I
Annonaceae	*Annona reticulata* L.	1	1	1	1	I
Annonaceae	*Annona squamosa* L.	1		1	1	I
Apocynaceae	*Nerium oleander* L.	1	1	1	7	I
Apocynaceae	*Plumeria obtusa* var. *sericifolia* (C. Wright) Woods	1	1	1	7	N
Apocynaceae	*Tabernaemontana alba* Mill	1		1	6	N
Apocynaceae	*Thevetia gaumeri* Hemsl. in Hook	1			7	N
Arecaceae	*Cocos nucifera* L.	1	1	1	1	I
Arecaceae	*Gaussia maya* (Cook) H. J.Quero & Read	1			7	N
Arecaceae	*Sabal mexicana* Mart	1	1	1	10	N
Arecaceae	*Veitchia merrillii* (Becc.) H.E.Moore	1		1	7	I
Bignoniaceae	*Crescentia cujete* L.	1			10	N
Bignoniaceae	*Parmentiera aculeata* (Kunth) L. O. Williams	1		1	1	N
Bignoniaceae	*Parmentiera millspaughiana* L. O. Williams	1			2	N
Bignoniaceae	*Tabebuia rosea* DC	1		1	2	N
Bixaceae	*Bixa orellana* L.	1			8	N
Boraginaceae	*Bourreria pulchra* Millsp. ex Greenm			1	2	N
Boraginaceae	*Cordia dodecandra* DC	1	1	1	1	N
Boraginaceae	*Ehretia tinifolia* L.	1		1	5	N
Cactaceae	*Nopalea cochenillifera* (L.) Salm-Dyck	1	1	1	6	I
Capparaceae	*Crateva tapia* L.		1		10	N
Caricaceae	*Carica papaya* L.	1	1	1	1	N
Combretaceae	*Terminalia catappa* L.	1	1		5	I
Euphorbiaceae	*Cnidoscolus aconitifolius* (Mill.) I. M. Johnst	1	1	1	11	N
Euphorbiaceae	*Jatropha gaumeri* Greenm		1	1	6	N
Euphorbiaceae	*Ricinus communis* L.	1			6	I
Fabaceae	*Vachellia farnesiana* (L.) Wight & Arn	1			10	N
Fabaceae	*Senegalia gaumeri* (S.F. Blake) Britton & Rose	1			4	N
Fabaceae	*Bauhinia divaricata* L.	1			6	N

(continued)

Table 4.1 (continued)

Family	Scientific name	R	Pu	Sr	Uses	OD
Fabaceae	*Bauhinia variegata* L.		1		7	I
Fabaceae	*Caesalpinia gaumeri* Greenm	1		1	4	N
Fabaceae	*Caesalpinia pulcherrima* Sw	1		1	5	N
Fabaceae	*Lonchocarpus punctatus* Kunth	1		1	2	N
Fabaceae	*Cassia fistula* L.			1	2	I
Fabaceae	*Diphysa carthagenensis* Jacq	1		1	2	N
Fabaceae	*Enterolobium cyclocarpum* (Jacq.) Griseb			1	5	N
Fabaceae	*Gliricidia sepium* Kunth ex Walp		1		9	I
Fabaceae	*Leucaena leucocephala* (Lam.) de Wit	1		1	9	N
Fabaceae	*Lonchocarpus yucatanensis* Pittier			1	2	N
Fabaceae	*Lysiloma latisiliquum* (L.) Benth	1			2	N
Fabaceae	*Mimosa bahamensis* Benth	1			4	N
Fabaceae	*Piscidia piscipula* (L.) Sarg	1		1	2	N
Fabaceae	*Havardia albicans* (Kunth) Britton & Rose	1			10	N
Fabaceae	*Chloroleucon mangense* var. *leucospermum* (Brandegee) Barneby & J. W. Grimes	1			2	N
Fabaceae	*Tamarindus indica* L.	1	1	1	1	I
Lamiaceae	*Gmelina arborea* Roxb. ex Sm			1	2	I
Lauraceae	*Persea americana* Mill	1	1	1	1	I
Malpighiaceae	*Bunchosia swartziana* Griseb	1			10	N
Fabaceae	*Myroxylon balsamum* (L.) Harms			1	10	N
Malpighiaceae	*Byrsonima crassifolia* Kunth	1	1	1	1	N
Malvaceae	*Ceiba pentandra* Gaertn	1			5	N
Malvaceae	*Guazuma ulmifolia* Lam	1		1	10	N
Malvaceae	*Malvaviscus arboreus* Cav	1	1		7	N
Malvaceae	*Hibiscus rosa-sinensis* L.	1	1	1	7	I
Meliaceae	*Azadirachta indica* A. Juss		1		6	I
Meliaceae	*Cedrela odorata* L.	1	1	1	2	N
Meliaceae	*Swietenia macrophylla* King	1		1	2	N
Moraceae	*Brosimum alicastrum* Sw	1		1	9	N
Moraceae	*Ficus carica* L.		1		1	I
Moraceae	*Ficus cotinifolia* Kunth		1		10	N
Moraceae	*Ficus máxima* Mill		1		7	N
Moraceae	*Chlorophora tinctoria* (L.) Gaudich. ex Benth			1	2	N
Morindaceae	*Morinda citrifolia* L.	1	1		6	I
Musaceae	*M+B28usa paradisiaca* L.	1	1	1	1	I
Myrtaceae	*Psidium guajava* L.	1	1	1	1	N
Myrtaceae	*Trichilia hirta* L.			1	10	N

(continued)

Table 4.1 (continued)

Family	Scientific name	R	Pu	Sr	Uses	OD
Phyllanthaceae	*Phyllanthus acidus* (L.) Skeels	1	1	1	1	I
Polygonaceae	*Gymnopodium floribundum* Rolfe	1			4	N
Punicaceae	*Punica granatum* L.	1	1	1	1	I
Rhamnaceae	*Karwinskia humboldtiana* (Willd. ex Roem. & Schult.) Zucc	1		1	2	N
Rhamnaceae	*Ziziphus* sp.				1	N
Rubiaceae	*Exostema mexicanum* A. Gray	1			10	N
Rubiaceae	*Hamelia patens* Jacq	1	1	1	6	N
Rubiaceae	*Rondeletia leucophylla* Kunth		1		7	I
Rutaceae	*Citrus aurantiaca* Swingle	1	1	1	1	I
Rutaceae	*Citrus aurantium* L.	1	1	1	1	I
Rutaceae	*Citrus* sp.	1	1	1	1	I
Rutaceae	*Citrus grandis* Osbeck	1		1	1	I
Rutaceae	*Citrus* × *latifolia* Tanaka ex Q. Jiménez	1	1	1	1	I
Rutaceae	*Citrus limettioides*Tanaka	1	1	1	1	I
Rutaceae	*Citrus limonia* Osbeck	1	1	1	1	I
Rutaceae	*Citrus paradisi* Macfad	1	1	1	1	I
Rutaceae	*Citrus medica* L.	1	1	1	1	I
Rutaceae	*Citrus reticulata* Blanco	1			1	I
Rutaceae	*Citrus sinensis* Osbeck	1	1	1	1	I
Rutaceae	*Murraya paniculata* (L.) Osbeck	1	1	1	7	I
Rutaceae	*Zanthoxylum caribaeum* Lam	1		1	6	N
Sapindaceae	*Melicoccus bijugatus* Jacq	1	1	1	1	I
Sapindaceae	*Talisia floresii* Standl			1	1	N
Sapindaceae	*Melicoccus oliviformis* Kunth ssp. *oliviformis*	1	1	1	1	N
Sapotaceae	*Chrysophyllum cainito* L.	1	1	1	1	N
Sapotaceae	*Malmea depressa* (Baill.) R.E.Fr			1	6	I
Sapotaceae	*Manilkara zapota* (L.) P. Royen	1	1	1	1	N
Sapotaceae	*Pouteria campechiana* (Kunth) Baehni	1		1	1	N
Sapotaceae	*Pouteria sapota* (Jacq.) H.E. Moore & Stearn		1	1	1	N
Simaroubaceae	*Bursera simaruba* (L.) Sarg	1		1	10	N
Solanaceae	*Cestrum diurnum* L.	1	1	1	7	N
Solanaceae	*Solanum erianthum* D. Don	1	1	1	7	N
Solanaceae	*Solanum hirtum* Vahl		1	1	6	N

Home gardens: R = rural. Pu = Peri-urban. Sr = Semi-rural. Uses: 1 = Fruit-tree. 2 = Timber. 3 = Fiber. 4 = Firewood or fuel. 5 = Shadow tree. 6 = Medicinal. 7 = Ornamental. 8 = Condiment. 9 = Forage. 10 = other uses. 11 = Edible OD (Origin distribution): N = Native. I = Introduced or nonnative

Source Based partly on Poot-Pool et al. (2015)

Fig. 4.3 Studies of peri-urban home gardens in the Yucatan peninsula, Mexico

the composition of that specific family garden. When a family no longer engages in agricultural work, but works in the services sector in a city; it adapts the composition and structure of the family garden to its new activity and economic strategy and

Fig. 4.4 Common species of urban home gardens in the Yucatan Peninsula, Mexico. Edible plants: A. *Mangifera indica* L. B. *Cocos nucifera* L. C. *Annona muricata* L. D. *A. reticulata* L. E. *A. squamosa* L. F. *Tamarindus indica* L. G. *Citrus limmetoides* Tanaka. H. *C. reticulate* Blanco. Ornamental plants: I. *Adenium obesum* (Forssk.) Roem. & Schult. J. *Nerium oleander* L. K. *Tabebuia rosea* DC. L. *Cassia fistula* L. *Photo credit* W. Cetzal-Ix

livelihood are completely different. The garden can then acquire the function of aesthetics by incorporating ornamental plants.

4.7 Conclusion

Home gardens from rural, semi-rural, urban as well as peri-urban perspectives are an important natural sanctuary of indigenous flora with unique combinations of species that help both the local community members in terms of better environment, aesthetics, income generation as well in restoration of vulnerable, local ecosystems, and protecting regional and local biodiversity. The urban home or family garden contributes significantly towards promoting local food security and better nutrition, help in strengthening family economy, processing of organic products, and in the conservation of regional and local agri-biodiversity.

Home gardens can successfully serve as a sustainable, green, environmental model practice for the environmental and ecological services, carbon sequestration, conservation and management of genetic diversity of plants and animals, provide shade and shelter for different threatened species such as urban wildlife, local bird and insect pollinators; and provide protection against natural disasters like hurricanes and accidental, anthropogenic fires from agriculture and livestock practices.

The home gardens from Mexico and other developing nations can serve as an efficient and effective model for sustainable ecosystem practices. Home garden approach can not only help in restoration of local vulnerable and degraded ecosystems and establishing and securing local biodiversity; but also help in adding much needed green breathing spaces to the urban environment that is now being overcrowded with concrete jungles. The concept and practices of family home gardens, can effectively help in reshaping urban ecosystems from the impact of pollution and help creating a better environment and a better tomorrow for our next generation.

References

Abdoellah O, Parekesit B, Hadikusumah H (2002) Home gardens in the upper citarum watershed, West Java: a challenge for in situ conservation of plant genetic resources. In: Watson JW, Eyzaguirre PB (eds) Home gardens and in situ conservation of plant genetic resources in farming systems. International Plant Genetic Resources Institute, Rome, Italy, pp 140–160

Abebe T, Wiersum KF, Bongers F (2010) Spatial and temporal variation in crop diversity in agroforestry homegardens of southern Ethiopia. Agrofor Syst 78:309–322

Akinnifesi FK, Sileshi GW, Ajayi OA, Akinnifesi AI, de Moura EG, Linhares JFP, Rodrigues I (2010) Biodiversity of the urban homegardens of São Luís city, Northeastern Brazil. Urban Ecosyst 13:129–146

Bernholt H, Kehlenbeck K, Gebauer J, Buerkert E (2009) Plant species richness and diversity in urban and peri-urban gardens of Niamey. Niger Agrofor Syst 77:159–179

CICENA (2017) Huertos escolares en México. Downloadable from http://www.ciceana.org.mx/

Cilliers SS, Matjila EM, Sandham L (2007) Urban agriculture utilizing the eco-circle approach in disadvantaged communities in Potchefstroom, South Africa. In: Stewart G, Ignatieva M, Bowring J, Egoz S, Melnichuk I (eds) Globalisation and landscape architecture: issues for references 121 education and practice. Polytechnic University Publishing House, St Petersburg, pp 88–91
Colding J (2007) Ecological land-use complementation for building resilience in ecosystems. Landscape Urban Plann 81:46–55
Eyzaguirre P, Watson J (2001) Home gardens and agrobiodiversity: an overview across regions. In: Watson JW, Eyzaguirre PB (eds) Home gardens and in situ conservation of plant genetic resources in farming systems. International Plant Genetic Resources Institute, Rome, Italy, pp 10–13
FAO (2010) Nueva política de Huertos escolares. Downloadable from www.fao.org/docrep/010/a1432e/a1432e00.htm
Fernández ECM, Nair PKR (1986) An evaluation of the structure and function of tropical homegardens. Agric Syst 21:279–310
Galluzzi G, Eyzaguirre P, Negri V (2010) Home gardens: neglected hotspots of agro-biodiversity and cultural diversity. Biodivers Conserv 19:3635–3654
Gaston KJ, Warren PH, Thompson K, Smith RM (2005) Urban domestic gardens: the extent of the resource and its associated features. Biodivers Conserv 14:3327–3349
Goddard MA, Dougill AJ, Benton TG (2009) Scaling up from gardens: biodiversity conservation in urban environments. Trends Ecol Evol 25:90–98
Heckler SL (2004) Cultivating sociality: aesthetic factors in the composition and function of Piaroa homegardens. J Ethnobiol 24:203–232
Kehlenbeck K (2007) Rural homegardens in central Sulawesi, Indonesia: an example for a sustainable agro-ecosystem?. Fakultät für Agrarwissenschaften Georg-August-University Göttingen, Germany, p 210
Lamont SR, Hardy-Eshbaugh W, Greenberg AM (1999) Species composition, diversity and use of home gardens among three amazonian villages. Econ Bot 53:312–326
Lubbe CS (2011) Comparison of the urban domestic garden flora along a socio-economic gradient in the Tlokwe City Municipality. Tesis de Maestría de la Universidad Master of Environmental Sciences at the North-West University, Potchefstroom Campus, p 175
Michon G, Mary F (1994) Conversion of traditional village gardens and new economic strategies of rural households in the area of Bogor, Indonesia. Agrofor Syst 25:31–58
Molebatsi LY, Siebert SJ, Cilliers SS, Lubbe CS, Davoren E (2010) The Tswana tshimo: a homegarden system of useful plants with a particular layout and function. Afr J Agric Res 5:2952–2963
Nemudzudzanyi AO, Siebert SJ, Zobolo AM, Molebatsi LY (2010) The Zulu muzi: a homegarden system of useful plants with a specific layout and function. Afr J Indig Knowl Syst 9:57–72
Novelo Chan V (2007) Influencia de la cercanía de la ciudad Valladolid, Yucatán a los huertos familares sobre el manejo de recurso vegetal. Bachelor Thesis, Faculty of Veterinary Medicine and Zoology, Autonomous University of Yucatan, Mérida, Yucatan, Mexico, p 63
Pauchard A, Aguayo M, Peña E, Urrutia R (2006) Multiple effects of urbanization on the biodiversity of developing countries: the case of a fast-growing metropolitan area (Concepción, Chile). Biol Conserv 127:272–281
Peyre A, Guidal A, Wiersum KF, Bongers F (2006) Homegarden dynamics in Kerala, India. In: Kumar BM, Nair PKR (eds) Tropical homegardens: a time-tested example of sustainable agroforestry. Springer, Dordrecht, pp 87–103
Pickett STA, Cadenasso ML, Grove JM, Nilon CH, Pouyat RV, Zipperer WC, Costanza R (2001) Urban ecological systems: linking terrestrial ecological, physical and socioeconomic components of metropolitan areas. Annu Rev Ecol Syst 32:127–157
Poot-Pool WS, van der Wal H, Flores-Guido S, Pat-Fernández JM, Esparza-Olguín L (2015) Home garden agrobiodiversity differentiates along a rural-peri-urban gradient in Campeche, México. Econ Bot 69:203–217
Rico-Gray V, García FJG, Chemas A, Puch A, Sima P (1990) Species composition, similarity, and structure of Mayan homegardens in Tixpeual and Tixcacaltuyub, Yucatan, Mexica. Econ Bot 44:470–487

SAGARPA (2017) Pone en marcha la SAGARPA y el SNTE el Programa "Mi Escuela Produce; parcelas escolares". Comunicación Social. Downloadable from http://www.sagarpa.gob.mx/Delegaciones/tlaxcala/boletines/Paginas/B0162017.aspx

Soemarwoto O, Conway GR (1992) The Javanese homegarden. J Farmin Syst Res Ext 2:95–118

Torquebiau E (1992) Are tropical agroforestry home gardens sustainable? Agr Ecosyst Environ 41:189–207

United Nations (2004) World urbanization prospects: the 2003 revision. Downloadable from http://www.un.org/esa/population/publications/wup2003/WUP2003Report.pdf. Accessed April 2017

Williams NSG, Schwartz MW, Vesk PA, Mccarthy MA, Hahs AK, Clemants SE, Corlett RT, Duncan RP, Norton BA, Thompson K, Mcdonnell MJ (2009) A conceptual framework for predicting the effects of urban environments on floras. J Ecol 97:4–9

Winklerprins AMGA (2002) Seasonal floodplain-upland migration along the lower Amazon river. Geogr Rev 92:415–431

Part II
Urban Landscaping and Design

Chapter 5
Toward the Creation of Urban Foodscapes: Case Studies of Successful Urban Agriculture Projects for Income Generation, Food Security, and Social Cohesion

Monique Centrone Stefani, Francesco Orsini, Francesca Magrefi, Esther Sanyé-Mengual, Giuseppina Pennisi, Nicola Michelon, Giovanni Bazzocchi and Giorgio Gianquinto

Abstract Urban agriculture (UA) is a strategic means of achieving sustainable urban food security now and into the future. While a number of different models of agriculture will be required to provide food for the concentrated number of people in urban spaces, UA can be key tool in helping urban residents meet micronutrient requirements, whose deficiency is now recognized as the most important cause of "hidden hunger" globally. UA, or the cultivation of crops for food in small and irregular spaces within a city and its periphery, has been practiced for as long as cities have existed. However, it is gaining increased attention for both its limited scale, its effects on nutrition and community building, the associated possibilities for employment and social mobility, its relevance in a vibrant and diverse food system and for all of these reasons, its sustainability. UA projects are springing up in cities all over the world, always engaging the collaboration of multiple urban stakeholders and increasingly with a focus on the most vulnerable people in society: the poor, the elderly, children, and those suffering malnutrition and chronic disease. For all city dwellers, UA improves livelihoods within the city environment by reducing city ecological footprint, strengthening city resilience to climate change and environmental hazards, reducing noise, improving air quality and the city microclimate, minimizing drought by improving storm water management, and contributing to solid waste management. It increases human health and well-being through stress

M. C. Stefani
Department of Sociology and Social Research, University of Trento, Trento, Italy

F. Orsini (✉) · E. Sanyé-Mengual · G. Pennisi · N. Michelon · G. Bazzocchi · G. Gianquinto
Research Center for the Urban Environment on Agriculture and Biodiversity (RESCUE-AB), Department of Agricultural Sciences (DipSA), University of Bologna, Bologna, Italy
e-mail: f.orsini@unibo.it

F. Magrefi
STePS s.r.l, Treviso, Italy

D. Nandwani (ed.), *Urban Horticulture*, Sustainable Development and Biodiversity 18, https://doi.org/10.1007/978-3-319-67017-1_5

recovery, encouragement of exercise, reduction of urban violence, facilitating social contact, and providing sources of added employment and income. Using the concept of urban political ecology as a foundation for understanding food dynamics, this chapter will describe a range of different UA initiatives across the globe, giving special attention to their multifunctionality in the context of social vulnerabilities within urban spaces. In so doing, it will present UA as an important platform for community engagement and sustainability building toward increased food security. Turning concrete into green productive lands results in changes in physical, psychological, environmental, and community health. At the same time, the application of social leftovers (abandoned lands, unused rooftops, plastic and wood containers, etc.) onto productive means creates a new model for adding micronutrition to urban diets that integrates and takes seriously social, cultural, nutritional, and economic outcomes.

Keywords Urban agriculture · Urban political ecology
Community engagement · Food security

5.1 Introduction

Food, and its production, is a complex part of social life, yet until more recently its political understanding remains relatively uni-dimensional, with a strong emphasis on national-level productive output. While some of the literature points to the importance of social factors in the creation of sustainable food security (Burchi et al. 2011), there has been little effort toward developing a richer approach that includes social factors in creating sustainability. Elaborating the meaning of food and nutrition security in its full complexity is necessary for promoting policies and programs toward *sustainable* food security. Even while such outcomes as improving dietary diversity or decreasing food insecurity, or increasing crop production may be explicit goals, one way to look toward sustainability is to focus on the myriad ways in which different urban populations benefit socially from growing food as an explicitly social activity, with social effects in addition to food- and nutrition-related outcomes. The following case studies demonstrate and provide lessons on how such activity may create generative value and engagement around food, by illustrating its integral connection to other important life activities—health and wellness, economic livelihoods, and social "support" for vulnerable groups such as immigrants and elderly. Tending to social outcomes as much as on productive capacity expands food and nutrition security from an explicit goal of increasing calories or micronutrients consumption, to multilevel nourishment within sustainable social activities, weaving it more fully into the lives of urban dwellers yet ultimately contributing to increased food security. Food as a product must be extricated from thinking about it as *solely* as a commodity, as this perspective masks the fact that cultivation activities have larger individual-, group-, and community-level effects. This reality is just as important within an urban setting that often values privacy and anonymity in the face of more concentrated population and that requires particular planning to maintain people's connection to nature.

Clearly, there are many intersecting strategies that can be used to promote increased urban food security and to reduce undernutrition, including strategic placement of retail outlets, subsidies, emergency relief services for vulnerable populations, school food programs, supplementation programs, and promotion of home gardening where possible. Urban agriculture (UA) is one way of approaching the problem of urban food insecurity, yet it is a potentially powerful way as it provides access to micronutrient-rich foods which marginalized regions in urban settings lack. Poor households tend to subsist on monotonous staple-based diets, lacking access to micronutrient-rich foods, such as fruits and vegetables (Burchi et al. 2011).

While a challenge of urban agriculture in Western cities is to some degree limited by the availability of land, the growth of urban farms in the United States has been in the most densely populated regions (Rogus and Dimtitri 2015). US food policy councils, local administrative boards that advise on policies to create fair and equitable food systems, have promoted urban agriculture (UA) as part of a repertoire of food security strategies. These signal a strong grass-roots response to changes in food access. Urban agriculture in this setting is viewed as a way to approach the problem of urban food insecurity directly and, more generally as a means to repair a broken food system as it has effects beyond food access itself. Within a larger food security strategy, urban agriculture is a powerful tool in how it provides more immediate access to micronutrient-rich foods and in how it may offer opportunities for locally based development through the implementation of urban agriculture projects.

Yet, while UA seems mainly about producing food, it does not take production per se as the cause of the problem of food insecurity. On the contrary, UA is as much about producing *access* to food as it is producing food, because of the very location of the activity—within urban centers where groups of residents lack fresh food access or do not have it within a reasonable proximity due to inadequate transportation or lack of retail outlets ("food deserts"). In fact, there are now a number of sources which indicate that while overall production levels are ample to feed everyone (Pimentel and Pimentel 2006), there is also a high proportion, of up to one-third of all food produced for human consumption that is wasted (FAO 2011) and that the location of food itself relative to people prevents greater levels of food security. This argument goes back to Amartya Sen's important analysis on the topic (Sen 1983).

In its contribution to food access, urban agriculture redefines the division of labor in food production between rural and urban spaces. As food requirements vary, for instance in times of war when needs may increase or be distributed differently, boundaries between urban and rural food production can also vary. Consistently, food security during WWII in many cities was ensured by UA, which was also promoted by national policies (e.g., "Dig for Victory" in UK, or the "Victory Gardens" in US). These policies filled a gap left by missing food supply from rural areas. Overall, the larger boundaries between rural and urban food production have been set by intense urban migration and urban development, which create greater economic value in urban property. Likewise, these boundaries are set within the commodification of agriculture, whereby technological advances such as advanced irrigation systems and other mechanization of agricultural process that allowed larger plots, are coupled with the concentration of ownership of agricultural land. The effect is a higher economic return to corporations through economies of scale that allow an accumulation of capital and power of corporations (McMichael 2011). Within the food system, these works to the disadvantage of vulnerable and food insecure urban stakeholders. While recognizing that food insecurity is partially based on poverty that limits economic access to food purchasing within a market-based system, access is overall based in such geographic and institutional organization which restrict options to inner city stakeholders within regional food systems. In this altered landscape, UA provides urban dwellers geographic access to improved diet. However, it also offers opportunities to participate in agricultural activities as a means to create more individual agency within the food system. Inequality in access to healthy food is due to a combination of conditions, a *social* ecology that includes geographic location (inner cities or marginal city spaces), poverty, and social inequality (in terms of the power of individuals to affect their environment), deeply intertwined with the physical or natural ecology—the land, the food, and the greater natural environment. Food access itself exists between the relationships of the social and the natural (Robbins 2012). The concept that seems to best capture such dynamics in the urban settings is called urban political ecology (UPE) (Agyeman and McEntee 2014).

The UPE approach employed to examine the following case studies takes as a baseline the history of geographic, social and institutional, economic and ecological conditions that create inequities in urban food systems and the means by which local food systems might find a new equilibrium. In this context, urban agriculture is viewed as a way to deal with urban food insecurity. Indeed, the main function of urban agriculture may be identified as its contribution to food and nutrition security, given that horticultural production inside cities enables the poorest to increase their access to food (Orsini et al. 2013) and addresses nutritional deficiencies (Kahane et al. 2005). According to USDA, about 15% of the world food is produced in urban areas (Wang and Nevius 2013). However, the analysis addresses aspects of how each of the projects are implemented within a local setting—with which organizations it connects or does not—and how it draws upon local resources for the development and maintenance of the program itself. Connecting with local stakeholders and the needs of both the environment and local people are seen as the core of a sustainable solution. Each project brings to urban agriculture a unique social emphasis, illustrating UAs resilience in adapting to a variety of contexts.

5.2 Economy of the Foodscape: Community Gardening Using Simplified Soilless Systems in the Suburbs of Teresina, Piaui, Brazil

Piaui is one of the poorest states of the Federal Republic of Brazil. Its capital city, Teresina, and the surrounding suburbs amount to more than 1,100,000 inhabitants, out of which more than 90% suffer from the scarcity of public infrastructure, health, water and sanitation services, or regular electrical supply. The absence of a waste collection service altogether with the nonexistence of a water sewage system, dramatically affects the incidence of disease (Campelo et al. 2005; Werneck et al. 2007). As a result, inhabitants are exposed to elevated risks of malnutrition, poverty and extremely poor life quality. Monthly income for many families is approximately 70 R$ (30 €). Similar to other cities in North-East Brazil, UA has been extensively promoted by institutional actors (the local city council, Prefeitura de Teresina, and research institutions involved in agriculture such as the Universidade Federal do Piaui and Embrapa) and approximately 52 allotment gardens can be found in urban marginal lands, covering an area of about 170 ha. In these gardens, individuals may obtain temporary contracts for use of a plot (renewed every four years) of approximately 400 m^2, where they receive both technical input and assistance from the local council to cultivate a range of vegetables (Monteiro and Monteiro 2006). Vegetables produced in these gardens are used both for self-consumption and for local sale. Earlier in the history of these plots however, gardeners' lack of technical skill, disaggregated product and seasonal fluctuations in production constrained product marketability thereby limiting potential gardeners' incomes from UA participation. Product quality was further reduced due to a high-gardeners turnover. With the aim of improving UA in Teresina, a community garden project was established in 2004 with financial support from FAO and the Italian Ministry of Foreign Affairs and the scientific guidance of experts from Italian and Brazilian universities (Fecondini et al. 2010). In this impoverished setting, UA was viewed as a factor that could improve urban livelihoods. This required changes in both production and market conditions for the gardeners. To reduce costs and improve vegetable's commercial quality, innovative growing systems were introduced using principles of simplified soilless cultivation, allowing a more intensive level of production. In order to match technology to the farmer's technical and organizational skills as well as to available materials, a consortium composed by both local and Italian Universities launched a pilot program with a new technical assistance center created specifically for this purpose (Casa da Hidroponia). By 2005, viable growing systems, substrates and nutrient solution were identified (Gianquinto et al. 2007), although the center is still conducting research and experimentation (Orsini et al. 2009, 2013) in order to provide the farming communities (organized as Clubes de Maes—mother's associations—involved in farming, each managing a garden) with state-of-the-art growing solutions and support in gardening design and management. Organizational changes that improved market conditions were also introduced with the start of the program in 2004. *Clubes de Maes* were grouped into a main farmers' association. Creating a cooperative association improved the gardeners' financial viability by ensuring regular customers for the gardeners' production. A fixed portion of the production is distributed (under specific terms and quantities) to city supermarket chains and restaurants. This arrangement is possible given the superior quality standards provided by better appearance of the produce, reduced seasonal fluctuations in production and the "social" value of a local fair-trade produce (Fecondini et al. 2009; Orsini et al. 2009). In order not to risk penalties due to missed delivery of the product, a portion of the production is given to the farmers association on a weekly basis for sale in street markets. At the same time, production from each *Clube de Maes* that is not absorbed by the farmers' association is either consumed by member families or locally sold (Fig. 5.1).

With external financial support having ended in 2005, the project has been maintained by nine gardener groups, representing about 107 households, who produce fresh vegetables (mainly lettuce) for the local market. Yearly production is estimated at 13,500 kg of lettuce, 2500 kg of coriander and 2500 kg of chives. Part of the production is reserved for institutional purchase by local nurseries and primary schools, where it is used for feeding about 20,000 school-aged children. Additionally, vegetables are distributed locally to about 15,000 people in vulnerable financial, nutritional, and social conditions. Overall, gardens are managed by the 120 members of the association "Grande Socopo Vegetable growers." As 75% of the associates are women, the project plays a crucial role in reducing gender inequality. About eight vegetable species (eggplant, coriander, cucumber, tomato, lettuce, okra, arugula, and bell pepper) that were previously almost absent in member's diets are today consumed on a daily basis by all family members, whose income has increased in measure from +25 to +100%.

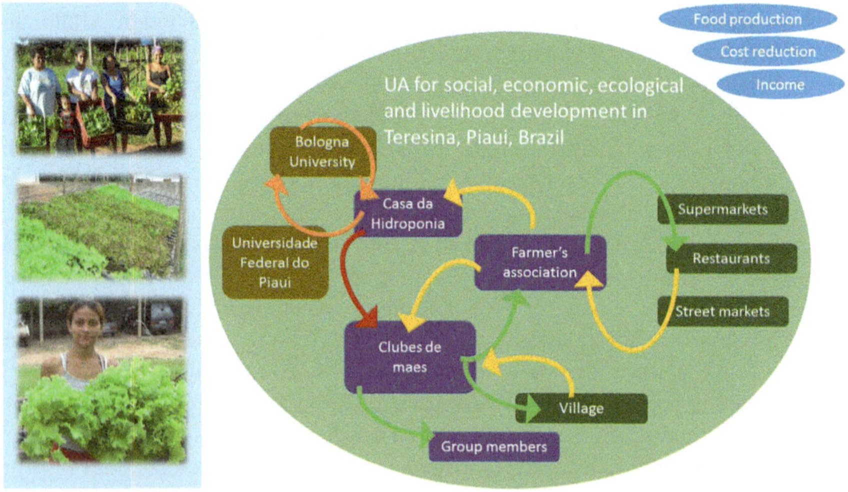

Fig. 5.1 Representation of the urban agriculture foodscape's economic dimension. Green arrows represent vegetable produce flow, in yellow the revenue, in orange the scientific guidance and in red the technical input and support

5.3 Health in the Urban Agricultural Foodscape: Assessing Relationships Between Food Security and Obesity in a Low-Income Population at the New Haven Farms

When urban households engage in community farming, consumption of more nutritious diets is generally observed (Zezza and Tasciotti 2010). Easy access to fresh agricultural goods can help to improve health conditions of the most disadvantaged urban dwellers. Yet a broken food system shows widespread population level effects. In 2011–2012, the adult population (>age 20) in the United States was estimated by the US Center for Disease Control to have an overweight rate of 69.0% and an obesity rate of 35.1%. In minority communities in the US, overweight/obesity rates are even higher than the general national average especially among Blacks and Hispanics where Hispanics are estimated at 75.7%. These obesity rates have been linked to less access to healthful, affordable food in the US. However, the obesity epidemic and associated chronic disease is considered a general result of the "nutrition transition" in many countries whose traditional diets are transformed to include cheaper, processed, and market-based foods (Popkin 2001). The occurrence of overweight and obesity in the US is prevalent among people of low socioeconomic status, who are those most likely to face food insecurity (Townsend 2006). Data for New Haven, Connecticut, and the surrounding county shows a relatively high rate of food insecurity at 13.6% although other local studies show the poorer sections of the town have up to 20% food insecurity (Fitzgerald et al. 2011). The sample selected for the New Haven Farms (NHF) project within a similarly poor section of the community for the NHF program had a rate of 66% food insecurity at baseline. According to NCHS data, New Haven ranks by a small margin as the second most obese county in the state at 26.7% and among the top two counties in the state with a diabetes rate of 8.3%. The "double burden" of people who are suffering diseases related to malnutrition at both ends of the curve is at the basis of several health problems suffered by poorest Hispanic residents in both the county and nationally.

In response to the experience of these problems as they were understood at a local health center in the city of New Haven, the founders of NHF Farm-Based Wellness program sought to address this problem through a food systems approach to reducing risk of diabetes. The nonprofit organization promotes health and community development through UA by establishing and cultivating urban farms that grow primarily vegetables. Individuals at risk of diabetes are referred to the program through their medical providers and are invited to take part in the 16–20-week program held during the growing season (approximately late June through October), that includes attendance at a weekly 2-hour session where they receive nutrition, cooking, and farm education and a weekly basket of vegetables produced at the farm. Educational sessions aim to increase knowledge of healthy eating: skills for planning and preparing balanced meals using the foods grown on the farm, as well as more basic healthy nutrition information. Each week the program focuses on foods available from the farm that week, so that members take home a free box of fresh vegetables with knowledge of the nutritional content of their food and new recipes that give them the foundation to improve their nutritional intake. The Farm-Based Wellness program aims to provide a holistic, educational approach to impact diabetes risk, exposing people to the origin of fresh food in a context of improving the community food system, greening urban spaces and giving urban residents at risk of chronic disease better access to fresh food.

NHF functions through a non-profit economic model, similar to the majority of small urban and peri-urban farms in the United States (Rogus and Dimitri 2015). Each season, this small program serves about 40–60 index members (those referred) and their families, totaling about 120 people, in a city of approximately 130,660 (USCB 2010). Its garden holdings include about one acre total of land in various parts of the city. The organization is masterful at creating community collaborations with myriad organizations including city offices, surrounding universities, and private businesses—an important form of community engagement. Donations of equipment and services from the surrounding community, including for example, soil donation, donation of properties held by both private and non-profit organizations as well as city-owned land register in value likely in the hundreds of thousands of dollars since the organization's founding in 2012 (where the acquisition of land remains one of the biggest challenges of urban agricultural projects). Donations, not including grants from various national, state and individual donors, comprise the majority of the operational budget. This deep network provides local-, small-, and medium-sized businesses visibility through their philanthropy. It also gives small business owners and other local donors in the community a stake in the success of the program, uniting them in the goal of community development and program achievement via better health and increased food security. Creating these community connections—which becomes the foundation of an enduring local structure—is part of building sustainability.

As it engages community organizations, NHF also engages individuals in the community in extensive volunteer work. The organization hosts only a handful of paid employees including the farmer, the executive director and a few other operational staff. The organization functions through the work of volunteers, which in many cases is equivalent to more than 15 hours per week. This network of volunteers, both individuals and organizations, demonstrates a crucial recognition and tangible support for the kind of activities NHF conducts in the community and for their health-related mission and outcomes. Within the program, NHF provided an important social connection (Fig. 5.2) for many of the members, those who managed to stay integrated within the program, with many reporting new social connections and friendships as well as other positive social- and health-related outcomes, including a significant reduction in food insecurity, both for program members and their families and a significant increase in the consumption of vegetables and to a lesser extent, whole versus processed grains.

In considering program sustainability and optimization, NHF might be seen as balancing two competing understandings of its own work. These competing goals may be impeding the creation of a more powerful program, one oriented more fully toward incorporating those served into the program work that could empower and ultimately generate greater community development as a result of its activities. The

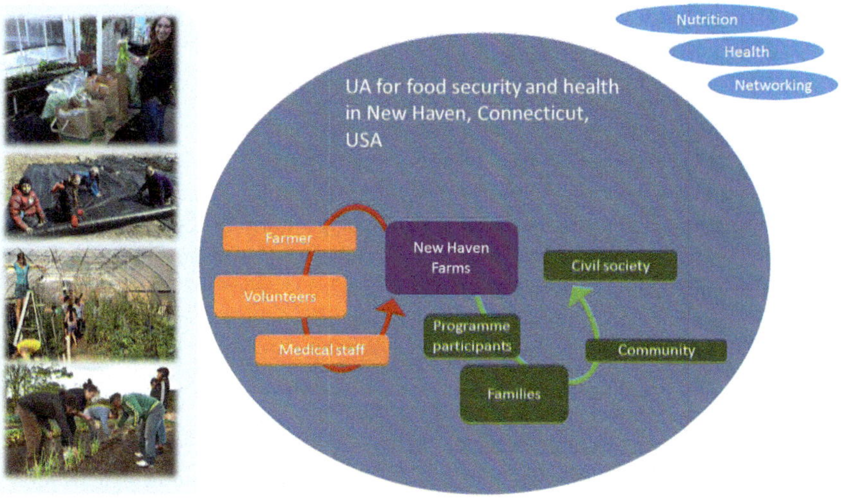

Fig. 5.2 Representation of the health features of urban agriculture foodscapes. Green arrows represent proper nutritional habits flow, in red the scientific guidance and technical input and support

first of these is to serve a food insecure and economically vulnerable population at risk for diabetes through exposure to local food, the consumption of fresh produce, and a food-based approach to health and chronic disease risk along with longer term food system change. A second and a more latent, unarticulated goal is the promotion of its program as a potentially scalable model whose activities could derive better funding by more important organizations (e.g., USDA and insurance companies). Ultimately, with much competition for grants and limited funding mechanisms for small studies with innovative approaches, the organization draws heavily upon a strata of volunteer professionals that are more institutionally trained, recognized, and settled than most of the vulnerable members of the community—those whom the organization serves. In this setting, the often impoverished members of the program do not benefit from direct participation in NHF programming itself. Even while farm education teaches people to grow their own food, the organization remains weak in empowering local individuals to replicate the program and, likewise, reproduces some of the dichotomies between those who have and do not have access to power and resources within the community.

5.4 Community Building in the Foodscape: Rooftop Community Gardening in Bologna, Italy

The city of Bologna has 382,784 inhabitants in 14,073 km^2. Bologna possesses a long history of urban horticulture initiatives (Orsini et al. 2014). Its municipal administration was among the first in Italy that regulated allotment gardens access and management. The main urban horticulture activities in the city are promoted by the City Council in collaboration with the National Association of Social Centres, Committees and Gardens of the Elders (ANCeSCAO) and other local social centers and civil society associations. The long history of UA in Bologna is accompanied by a vibrant program and technical expertise that has developed in the University of Bologna, which has consulted on and helped to build UA in numerous projects and programs around the world (Orsini et al. 2013).

In the historical center of Bologna, urban gardening derives primarily from a unique architecture and city planning which dates back to the middle age. Several of these gardens may still be encountered in the center of Bologna, but there has been growth in gardening activity in peripheral areas as well (Gasperi et al. 2016). The city is structured as a network of green infrastructures (e.g., public parks and gardens), that can be connected together by a set of greened streets, porches, and courtyards. Historically, internal gardens (called *Hortus conclusus* or "enclosed garden") were developed in private courtyards as open spaces enclosed by fruit trees and small crops irrigated by water channels (Scannavini and Palmieri 1990). On this foundation, urban agriculture was incorporated as a socially, architecturally, and politically inherent part of the landscape of the city center. However, beginning in the mid-twentieth century (1951–1971), industrialization and rapid population growth and migration to large cities, stimulated by the Marshall Plan after WWII, relegated urban agriculture to the marginal areas of the city (beside main roads, for instance, which is more common in the developing world), generally occupied without authorization from the city council (Crespi 1982). During the same period, UA as an ongoing and institutionalized civic activity slowed down dramatically as intensive marketing on the quality and safety of industrially prepared food grew.

In 1982 *Italia Nostra* (a National non-profit body devoted to the preservation of historical and natural heritage), defined urban garden as "individual plots of a few hundred square meters cultivated directly by the holder and his family, whose product is used predominantly for self-consumption and takes part in a cycle of barter economy." This effort attempted to bring attention of local administrations to the increasing relevance of urban gardening. At the beginning of the 80s, around 60 ha of urban gardens could be found in Bologna. Most of the plots were managed by the city council and were located in new public housing development areas often for use and occupied by the elderly (Crespi 1982). These years, gardeners were mainly people engaged in repetitive work activities, who found an ideal leisure activity and distraction within urban gardens.

In Bologna today there are 20 horticultural areas and an estimated 3000 urban gardens over 100,000 m^2 (each composed by 40 to 300 plots each of about 40 m^2). The majority of gardens are represented by allotment and school gardens. Moreover, due to long waiting lists for space in existing community gardens held by the city, there are currently a number of urban garden sites developing along roadways, railways and waterways around the periphery of the city, but also within residential buildings.

Concurrently, there is a developing trend toward community managed gardening initiatives. Among them, the first social housing rooftop garden (in Via Gandusio, 10) was implemented in 2010, based upon a collaboration of the Bologna city council, the Bologna University and a non-profit organization Biodiversity (Orsini et al. 2014) (Fig. 5.3). The project, called "GreenHousing," addresses a need to create a greener and more sustainable city and improve relationships between people. The stated objectives from the beginning were *sustainable cultivation management*, achieved through avoidance of pesticides, the promotion of recycling and reuse of materials, as well as the preservation of natural resources (e.g., water) and traditions (e.g., seeds, varieties). Moreover, because of the unique rooftop location, the adoption of innovative techniques that could bypass the absence of a soil base was crucial. The garden is located at the eleventh floor of a gray cement building created in the 1970s as a public housing initiative and is today inhabited by about 300 people in 120 living units. The structure had received little maintenance in recent years and conflicts have been prevalent among inhabitants, which are mainly composed of elderly Italians and younger immigrant families. Since the revival of urban agriculture in the late twelfth century, there has been a general increase of immigrants in urban gardens. In part, this is due to a general increase of people immigrating to the city. In Bologna, more than

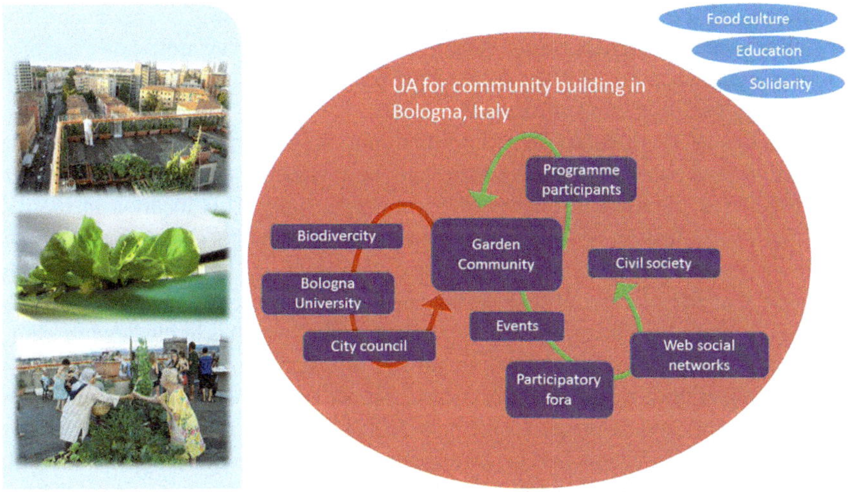

Fig. 5.3 Representation of the social features of urban agriculture foodscapes. Green arrows represent social relations while in red the scientific guidance and technical input and support

52,000 foreign people live in the municipal area, about 13.5% of the urban population. Nevertheless, immigrants' presence in UA activity is also due to the greater level of food insecurity and as means for social networks and incorporation. After initial difficulties that alternated between disregard or complaints from the residents who considered other types of intervention of greater priority, the project formed a garden management group from a small number of residents and has become self-sustaining with technical assistance and social support continuing from the University. The garden does not satisfy all food needs for the core members and their families, about 25 people in total. However, it allows them a richer diet and in the most productive months of the year, enough so that food is distributed to enrich the diet of another 10–15 resident families in the building.

5.5 Discussion

McClintock (2010) rightfully points to the distinction that is often neglected when discussing urban agriculture: that its uses and practices will differ between Global North and South. There are clear differences that exist, such as the size of the plots available. This difference is reflected in the sample of projects include here. At the same time, urban agriculture does provide both food and social connections that create forms of resilience for people in poverty and other social groups that merits the comparative examination of such projects, despite geographic location. For instance, while the differences may be between "limited," "inadequate" or "non-existent" access between global regions, poor people in urban areas in the US that are characterized by racial segregation and income inequality suffer from lack of fresh food access as they tend to have food sources usually a mile away or greater and have no or little access to transportation (USDA 2009). In these same areas however, there is normally an abundance of fast food which puts people at high risk for overweight and obesity and the chronic disease that results.

The plots of all three projects described here were located in areas of the city that are impoverished. In the Gandusio and New Haven Farms projects, members consisted of a majority of older people and immigrants. This location of the plot facilitates participation of local people and brings the benefits of UA directly to the neighborhood of the participants. Likewise, anecdotal evidence from participants suggests that the placement and participation in the gardening activities offers a feeling of ownership and pride in project participation, which is likely key in creating sustainability at the level of the individual and organization.

The zone of the city in which gardening plots are located will have an effect on the size of the plots and therefore on the possibility of sharing resources as well as the ways that urban gardeners can benefit from gardens. In the US, Rogus and Dimitri (2015) have shown that the growth has been greatest in areas of greater land value, indicating that plots will likely be smaller. This will radically limit the possibility for urban gardeners in the US to earn a large enough margin to support families. Areas where larger plots (such those of few hundred meters that can be found in

Teresina allotment gardens) can be created to allow a greater level of organization and coordination and also a supplemental income. Indeed, part of the turnover of the Teresina gardens has been that gardeners' income generation becomes a motive for ultimately leaving the area to obtain better work and housing. However, in the New Haven Farms program, a high proportion of members (>50%) indicated that the food provided at the farm for the 20 weeks during the growing season allowed them to reallocate resources normally spent on food to other nonfood purchases. This indicates that even when garden participation will not bring an income, it adds to overall resources in the household.

In all of the gardens, the consumption of vegetables increased within the diets of participants. In the case of New Haven Farms, whose mission is health related, the organization had more rigorous measures of dietary intake, using a national validated survey in English and Spanish, the Dietary Screener of the National Health Administration Nutrition Examination Survey. In this program, the survey showed that the average increase in vegetable intake was nearly one-serving a day for index members. Although the survey did not cover all household members, it is likely that their intake increased as well. Each of the programs reported increase in vegetable intake although without more precise quantification.

This brings us to one of the more notable findings of the comparison of urban farming programs to vulnerable social groups. In each of the programs, participants needed to be willing to participate, but were not explicitly seeking gardening activities, nor were they trying more specifically to cope with inadequate food resources, for instance at a food bank. Measuring intake on many of the urban farms projects in the US has tended to focus on those who are self-selecting into gardening activities and are therefore more likely to show increased intake as a result of participation (Wakefield et al. 2007; Litt et al. 2011). These projects illustrate that participation in UA for groups vulnerable to food insecurity will experience dietary benefits while not explicitly seeking them or without having articulated motivations to make dietary changes.

Additionally, many initially had no knowledge of gardening techniques. They were ultimately connected to the programs through the promotion of health and wellness, through opportunities for greater income and for increased social opportunities and social support. Yet as socially vulnerable groups, a high proportion suffered from food insecurity. Each of the programs offered to a greater or lesser degree, food, technical assistance, and nutrition information that ultimately seemed to allow participants to increase food security and corresponding dietary intake of fresh food. Targeting *social* groups vulnerable to food insecurity and engaging them in gardening activities within the context of a larger social support system, pulls food insecure people that were unlikely to seek out such opportunities into a process that ultimately offers some multiple forms of relief.

It should be noted however that merely placing farmers markets in the city centers is unlikely to have the same effect. When participants grow the food themselves, they have even more direct access than even good market proximity affords because it eliminates the financial barrier. Farmer's markets, at least in the US have been noted for potentially reproducing inequalities as they target a higher socioeconomic group (Golden 2011). Moreover, there are greater benefits that participation in an active work environment offers—sociability, new knowledge acquisition and network acquisition.

There are a number of differences in the way each of the programs were organized and administered and in the role gardeners played in the overall program. In the case of New Haven Farms, the organization is structured as a nonprofit organization providing a program with a health mission. Although program participants took part in didactic activities such as weeding and harvesting, they did not perform the bulk of the work, which was done by staff and volunteers. Program participants, were exactly that—"participants" as opposed to the "gardeners" of the other programs. In this way, the NHF program might be considered more of a traditional medical program, with participation as a treatment provided by staff who delivers a service. In this sense, program adherence in this model is a challenge in the same way medication adherence can be, and it overall has limitations in the extent to which participants are empowered as a result.

It is clear in all cases that access to land with public agreements allows for a more coordinated response from collaborating organizations, civic authorities, university or other technical expertise as well as funding than does sites that are set up in peripheries without any governing bodies involved. At the same, it should be emphasized that it is the legal, financial, educational, and institutional resources are there to provide support to the community as a project is initiated. However, if the goal is to create sustainability, then it should be considered mandatory that members of the vulnerable communities or representatives be part of the planning and organizing of gardens. That is, rather than consider the gardens as "service providers," they should be understood as community development. As such, the knowledge developed or advanced in the course of organizing such projects should not be considered solely within the domain of project leaders; knowledge transfer in all domains including management, accounting and administration is key in empowered community development as the Teresina garden illustrates. In each of the cases presented here, the goal was mainly to transfer knowledge of nutrition or gardening techniques. The more, most inclusive and empowering transfer of knowledge would also include project management, long-term planning and simple forms of program evaluation.

5.6 Conclusion

Supporting urban agriculture is an important measure that city governments can take toward the problem of urban food insecurity. A comprehensive means of understanding the history and the combination of factors affecting the food circumstances of those in urban environments is revealed through the urban political economy perspective. Engaging groups based on social vulnerabilities as opposed to those who are already known to be food insecure in urban gardening activities ensures that programs and projects targeting food insecure will select in new individuals. In the absence of solid community or regional level data on food insecurity, a problem sounded increasingly by leadership in the nutrition science community, such measures allow multiple community needs to be satisfied simultaneously.

Acknowledgements For the useful information on the study cases presented in this manuscript, we wish to acknowledge Elizabeth Marsh, Daniela Gasperi and Luana Iori.

References

Agyeman J, McEntee J (2014) Moving the field of food justice forward through the lens of urban political ecology. Geog Compass 8(3):211–220

Burchi F, Fanzo J, Frison E (2011) The role of food and nutrition system approaches in tackling hidden hunger. Int J Environ Res Pub Health 8:358–373

Campelo V, Gonçalves MAG, Donadi EA (2005) Mortalidade por doenças infecciosas e parasitárias no Município de Teresina-PI (Brasil), 1971–2000. Rev Bras Epidemiol 8(1):31–40

Crespi GC (1982) Orti urbani: una risorsa. Franco Angeli Editore, Milano

FAO (2011) Global food losses and food waste—extent, causes and prevention, Rome

Fecondini M, Casati M, Dimech M, Michelon N, Orsini F, Gianquinto G (2009) Improved cultivation of lettuce with a low cost soilless system in indigent areas of north-east Brazil. Acta Hort 807:501–507

Fecondini M, Damasio de Faria AC, Michelon N, Mezzetti M, Orsini F, Gianquinto G (2010) Learning the value of gardening: results from an experience of community based simplified hydroponics in north-east Brazil. Acta Hort 881:111–116

Fitzgerald N, Hromi-Fiedler A, Segura-Pérez S, Pérez-Escamilla R (2011) Food insecurity is related to increased risk of type 2 diabetes among Latinas. Ethn Dis 21(3):328–334

Gasperi D, Pennisi G, Rizzati N, Magrefi F, Bazzocchi G, Mezzacapo U, Centrone Stefani M, Sanyé-Mengual E, Orsini F, Gianquinto G (2016) Towards regenerated and productive vacant areas through urban horticulture: lessons from Bologna, Italy. Sustain 8(12):1347

Gianquinto G, Orsini F, Michelon N, Ferreira da Silva D, Damasio de Faria F (2007) Improving yield of vegetables by using soilless micro-garden technologies in peri-urban area of North-East Brazil. Acta Hort 747:57–65

Golden S (2011) Urban agriculture impacts: social, health, and economic: a literature review. In: University of California sustainable agriculture research and education program agricultural sustainability institute at UC Davis

Kahane R, Temple L, Brat P, de Bon H (2005) Les légumes feuilles des pays tropicaux: diversité, richesse économique et valeur santé dans un contexte très fragile. In: Parrot L, Njoya A, Temple L, Assogba-Komlan F, Kahane R, Ba Diao M, Havard M (eds) Agricultures et développement urbain en Afrique subsaharienne: Environnement et enjeux sanitaires, Editions L'Harmattan, pp 119–129

Litt JS, Soobader M-J, Turbin MS, Hale JW, Buchenau M, Marshall JA (2011) The influence of social involvement, neighborhood aesthetics, and community garden participation on fruit and vegetable consumption. Am J Pub Health 101(8):1466–1473

McClintock N (2010) Why farm the city? theorizing urban agriculture through a lens of metabolic rift. Cambridge J Reg Econ Soc 3:191–207

McMichael PD (2011) Food system sustainability: questions of environmental governance in the new world (dis)order. Glob Environ Change 21:804–812

Monteiro JPR, Monteiro MSL (2006) Hortas comunitárias de Teresina: agricultura urbana e perspectiva de desenvolvimento local. Rev Iberoamericana Economía Ecol 5:47–60

Orsini F, Michelon N, Scocozza F, Gianquinto G (2009) Farmers-to-consumers: an example of sustainable soilless horticulture in urban and peri-urban areas. Acta Hort 809:209–220

Orsini F, Kahane R, Nono-Womdim R, Gianquinto G (2013) Urban agriculture in the developing world. a review. Agron Sust Dev 33:695–720

Orsini F, Gasperi D, Marchetti L, Piovene C, Draghetti S, Ramazzotti S, Bazzocchi G, Gianquinto G (2014) Exploring the production capacity of rooftop gardens (RTGs) in urban agriculture: the potential impact on food and nutrition security, biodiversity and other ecosystem services in the city of Bologna. Food Secur 6:781–792

Pimentel D, Pimentel M (2006) Global environmental resources versus world population growth. Ecol Econ 59(2):195–198

Popkin B (2001) The nutrition transition and obesity in the developing world. J Nutr 131(3):871S–873S

Robbins J (2012) Diet for a new America 25th anniversary edition: how your food choices affect your health, your happiness, and the future of life on earth. HJ Kramer

Rogus S, Dimitri C (2015) Agriculture in urban and peri-urban areas in the United States: highlights from the census of agriculture. Renew Agr Food Syst 30:64–78

Scannavini R, Palmieri R (1990) La storia verde di Bologna: strutture, forme e immagini di orti, giardini e corti. Nuova Alfa, Bologna

Sen A (1983) Poverty and famines: an essay on entitlement and deprivation. Oxford University Press

Townsend MS (2006) Obesity in low income communities: prevalence, effects, a place to begin. J Am Diet Assoc 106(1):34–37

United States Census Bureau (2010) Census, U.S

United States Department of Agriculture (USDA) Economic Research Service (2009) Access to affordable and nutritious food: measuring and understanding food deserts and their consequences. Washington, DC

Wakefield S, Yeudall F, Taron C, Reynolds J, Skinner A (2007) Growing urban health: community gardening in South-East Toronto. Health Prom Int 22(2):92–101

Wang PP, Nevius JG (2013) Promoting urban agriculture: insurance as an important tool. Environ Claims J 25(4):320–330

Werneck GL, Costa CHN, Walker AM, David JR, Wand M, Maguire JH (2007) Multilevel modelling of the incidence of visceral leishmaniasis in Teresina. Brazil Epidemiol Infect 135(2):195–201

Zezza A, Tasciotti L (2010) Urban agriculture, poverty and food security: empirical evidence from a sample of developing countries. Food Policy 35:265–273

Chapter 6
Smart Precision Lighting for Urban and Landscape Closed Controlled Horticultural Environments

D. Piromalis, K. G. Arvanitis, P. Papageorgas and K. P. Ferentinos

Abstract Although the benefits of stimulating and control of the various physiological properties and growth of plants through the use of artificial lighting have been proven during the last decades, artificial lighting is facing many challenges today, especially due to the fact that horticulture has become a rapidly changing sector. Closed controlled horticultural environments go beyond the boundaries of the established professional greenhouses and move to various urban deployments. On the other hand, there are significant technologies, either from other established application domains or new challenging ones, that can be beneficially integrated into the existing lighting systems for horticulture. In this study, the particular requirements of closed controlled horticultural environments related to the artificial lighting are identified and presented categorized in functional, practical, electrical, and economical requirements. Moreover, the technological context through which the smart precision lighting applications can be met is introduced. The most common of the traditional lighting sources used in horticulture are reviewed and the new revolutionary technology of the solid-state lighting (SSL) and its advantages for horticulture is exhibited. Additionally, certain SSL fixtures design cases, as a proof of the design options in SSL fixtures, are also presented. Subsequently, the energy management

D. Piromalis (✉)
Department of Industrial Design and Production Engineering, University of West Attica, Egaleo, Athens, Greece
e-mail: piromali@puas.gr

K. G. Arvanitis
Department of Natural Resources Management and Agricultural Engineering, Agricultural University of Athens, Athens, Greece

P. Papageorgas
Department of Electrical and Electronics Engineering, University of West Attica, Egaleo, Athens, Greece

K. P. Ferentinos
Department of Agricultural Engineering, Institute of Soil and Water Resources, Hellenic Agricultural Organization "Demeter", Athens, Greece

D. Nandwani (ed.), *Urban Horticulture*, Sustainable Development and Biodiversity 18, https://doi.org/10.1007/978-3-319-67017-1_6

approaches in SSL deployments are explained, the available networking technologies for interconnected lighting fixtures are reported and comments are given on the current commercial approaches, while the challenging concept of the networking of clusters of lighting fixtures is reviewed. The associated advantages for smart precision lighting in horticulture are also reported. Finally, the study focuses on the development of intelligence in lighting control.

Keywords Solid-state lighting · Artificial lighting · Urban horticulture · Wireless sensors networks · Precision lighting · LED lighting · Greenhouse lighting

6.1 Introduction

Natural light, generated by the solar radiation during the day, is one of the most critical factors for plants growth. Sunlight is considered as an electromagnetic radiation that is comprised of three distinct wavelength zones. The first zone includes the ultraviolet spectrum from 200 up to 380 nm, and it is not visible to humans. The second zone includes the visible spectrum from 400 up to 720 nm which includes the bands of violet, blue, green, orange, and red colors. The third zone includes the non-visible far-red and infrared spectrum from 720 up to 1000 nm. Any wavelength over 1000 nm is solely converted into heat. Even though human vision perceives sunlight as white light, in practice this is comprised of a mix of colors in various intensities.

On the other hand, plants have a totally different perception to the natural light than humans. Specifically, they strongly respond to wavelengths that blue and red colors are dominant, in order to intercept photon flux by their chloroplast cells, and convert carbon contained to the air into carbohydrates. This process, known as *photosynthesis*, is mainly responsible for the growth of the plants. The effects of natural light on the plants have been studied for many decades, with important results. The knowledge of the plants inherent mechanisms for managing lighting is very useful for understanding the crops' development patterns associated with particular weather conditions and latitudes in open field cultivations. Nevertheless, the real benefits are on the side of cultivation systems inside closed controlled environments. In these environments, crops growth and harvesting phases are controlled and not governed by the seasonally natural environmental conditions. Attempting to take the full potential of natural lighting in closed controlled environments, several approaches and techniques have been developed regarding the construction materials and the architectural characteristics of greenhouses (Ponce et al. 2014).

Unfortunately, the daily average flux of the solar radiation is not constant in a yearly basis. For this reason, and in an effort to maximize the quantity and quality of crops production, researchers intensified their studies on the plants behavior to artificial lighting (Johansen et al. 2011; Vänninen et al. 2010). Thus, today there is plenty of information about the plants' response to the light spectrum. As illustrated in Fig. 6.1, plants response varies to different wavelength bands of the solar radiation spectrum.

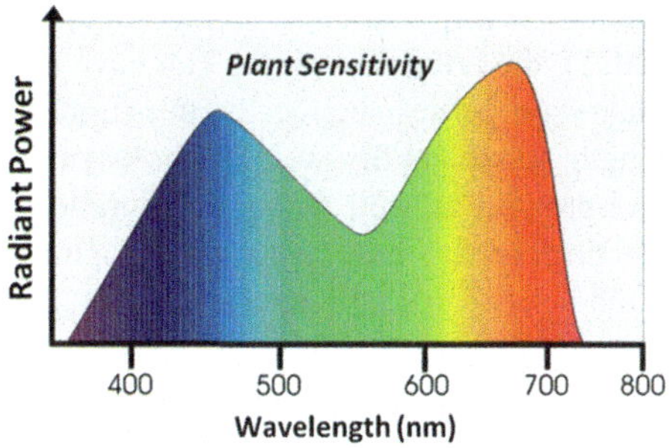

Fig. 6.1 Plants' sensitivity to solar radiation spectrum

Based on this knowledge, many studies have been performed to correlate the impact of light to significant plants processes and mechanisms, such as, photosynthesis, photoperiod growth, germination, vegetative growth, budding, photomorphogenesis, evapotranspiration, and abnormal development processes such as etiolatia. In summary, the graph in Fig. 6.2 presents how specific light wavelengths influence particular plants' processes (OSRAM 2015). According to this graph, photosynthesis is stronger in the wavelengths of blue and red lights, while photoperiod growth and germination are stronger in the wavelengths of red (hyper-red) a far-red (infrared) lights.

During the last few decades, various specific plants studied regarding the influence of light on them. Typical cases can be found for many of the crops cultivated in closed controlled horticultural environments, i.e., vegetables (tomatoes, cucumber, etc.), and ornamental plants like cut flower (chrysanthemums, roses, etc.), and potted plants

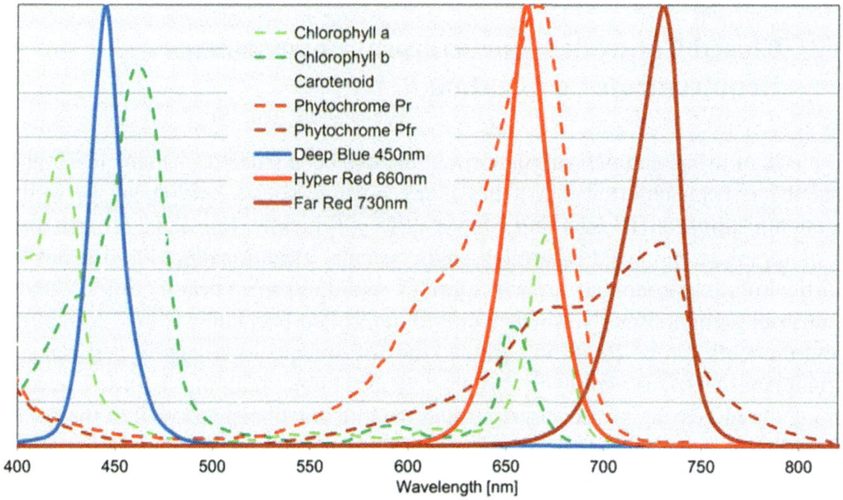

Fig. 6.2 Plants' response to specific wavelengths of light

(dendrobiums, bromelias, orchid/Phalaenopsis, etc.). As it is reported by Singh et al. (2015), the artificial lighting can have a positive influence in the control of certain plants' properties such as, germination, flowering, leaves shapes, bracts and stems length, fresh and dry weight, photosynthetic pigment content, water requirements, and growth, as well as harvest duration. Today it has been proven that artificial lighting entails significant benefits for horticulture in closed controlled environments (Böhme et al. 2015; Gömez and Mitchell 2016; Han et al. 2015), but in order to meet the complete requirements of specific cultivations and to allow the realization of smart precision lighting applications towards the twenty-first century, it is necessary to integrate certain new technologies.

This study is comprised of six sections. In Sect. 6.2, the particular requirements of closed controlled horticultural environments related to the artificial lighting are identified and presented categorized in functional, practical, electrical, and economical requirements. Also, Sect. 6.2 introduces the technological context through which the smart precision lighting applications can be met. Section 6.3 reviews the most common of the traditional lighting sources used in horticulture and it presents the new revolutionary technology of the solid-state lighting (SSL) and its advantages for horticulture. Additionally, Sect. 6.3 presents certain SSL fixtures design cases as a proof of the design options in SSL fixtures. Section 6.4 first explains the energy management approaches in SSL deployments, second it presents the available networking technologies for interconnected lighting fixtures, and comments on the current commercial approaches, and third it reviews the challenging concept of the networking of clusters of lighting fixtures and explains the associated advantages for smart precision lighting in horticulture. Section 6.5 focuses on the development of intelligence in lighting control.

6.2 Closed Controlled Horticultural Environments Requirements for Lighting

The role of urban and peri-urban horticulture has been studied by many researchers and it has been reported that it has many positive social, ecological, as well as economic impacts (Ulrichs and Mewis 2015). Typically, this kind of agriculture is taking place in closed controlled environments. Traditionally, closed controlled horticultural environments are synonymous with the production facilities found in industrial greenhouses. Nowadays, closed controlled horticultural and floricultural environments can be found in various types of botanical gardens in peri-urban or urban landscapes, as well as in various civil and public building facilities. Furthermore, the increasing awareness regarding the climatic change as well as the quality of food, has urged individuals to be involved with urban horticulture in various forms of internal farming. For instance, internal farming can be met in the form of small gardens for vegetables, micro-greenhouses in alleys, plant factories, vertical farming inside buildings, etc.

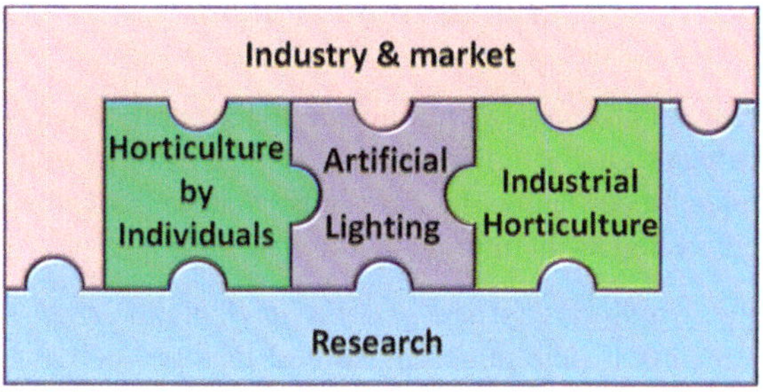

Fig. 6.3 The multi-facet and complex context of requirements for artificial lighting systems in closed controlled horticultural environments

On the other hand, in a race for continuous increase of food production, the boundaries of the closed controlled environments in greenhouses are continuously changing. Initially from soil to soilless techniques (*hydroponics*), and now to cultivations without any substrate for rooting system support, as in *aeroponics*. The concept of *vertical farming* will flourish in the twenty-first century, provided that there would will be a collaboration among various conglomerate experts, such as architects, engineers, urban planners and others, with the horticulturists (Eigenbrod and Gruda 2015). Hence, the meaning of the closed controlled environments in horticulture as well as floriculture varies strongly according to the particular aforementioned applications. In addition, there are many differences among various existing implementations in terms of crops species, the scale of cultivations, the sense of profit, the investment cost, the idiosyncrasy of the end users and consumers, and so on. At the same time, there are also several common things among them, with artificial lighting probably being the most critical one (Kozai and Niu 2015; Takagaki et al. 2015).

Taking into consideration the importance of urban horticulture for the food security and the sustainability of modern cities in both the developed and the developing countries worldwide on one hand, and the transformation of cultivations in regards to the variety of closed controlled environments on the other hand, artificial lighting should be able to fulfill several requirements of different stakeholders. As it is illustrated in Fig. 6.3, artificial lighting is placed in the middle of a complex and intercorrelated context of groups with different vested interests on horticulture, such as the group of individuals which are growing their own plants in customized closed controlled environments, the professional growers which operate their cultivations in industrial-grade greenhouses, the industry and market of systems, raw materials, resources, and horticultural products, and the researchers from horticulture and other scientific domains. The requirements for artificial lighting in horticulture can be grouped into four distinct categories, namely the functional, the practical, the electrical, and the economical requirements.

6.2.1 Functional Requirements

Functional requirements include all the purely associated with the function of lighting in regards to the lighting quality and intensity, as well as with the lighting systems features. In particular, the most important functional requirements identified for closed controlled horticultural environments are

(a) *Right quality and quantity of light*: the right spectrum and intensity of light flux is determined by several factors, such as the particular plant species, the average daylight duration, latitude of the cultivation, and weather conditions that influence sunlight due to different angle of the sun, architectural and building characteristics of the facilities that probably influence incoming daylight, and the most important in terms of yield, the specific physiological properties of the plants that need to be affected.

(b) *Right lighting operation mode support*: according to the requirements of the specific plant species, artificial lighting systems have to occasionally support one of three distinct operation modes (OSRAM 2015): the *replacement lighting mode*, under which the lighting systems are the sole source of lighting in cases, e.g., such as closed controlled growing rooms and growth chambers (Contreras et al. 2008); the *supplemental*, or else, the *production lighting mode*, which is very common in cases where the average daily sunlight is not sufficient to meet the lighting requirements of the growing plants; and, the *photoperiod lighting mode*, which is used to stimulate or influence the photoperiod-dependent responses of the plants, such as the flowering of ornamentals and the growth of vegetables (Gömez and Mitchell 2016).

(c) *Spatiotemporal variability*: the artificial lighting systems have to provide operating functions for time programming in order to satisfy the fluctuations of the daylight in a yearly basis. In other words, the duration of the lighting systems should be fully time-defined in the span of the day, the week, and the year.

(d) *Flexibility and reconfigurability*: flexibility stands for the required function of the lighting system to be occasionally reprogrammed in order to execute different application scenarios in a yearly basis. Furthermore, the lighting systems have to be reconfigurable. Specifically, the lighting fixtures that comprise the artificial lighting system should be able to be deployed in clusters so as to support different growing application scenarios, e.g., multiple different cultivations in the same closed and controlled area. The reasoning behind the clustering patterns has to be determined according to several similarities amongst the lighting cluster members, such as the chromaticity or the physical positioning of particular cultivations.

(e) *Environmental variability*: it is necessary for the lighting systems to respond to direct disturbances such as sunlight variation, as well as to indirect disturbances such as temperature, humidity and carbon dioxide concentration that correlate to the evapotranspiration of the plants.

(f) *Controllability*: the lighting system should be adjusted according to the particular requirements of the horticultural application in terms of light spectra and density. Ideally, the very same lighting system should be used for a variety of applications that take place in the same environment during the various crops seasons of the year.

(g) *Automation and supervision*: ideally, the lighting system should not be considered as an independent system, but as an integral part of the overall automation system of the closed controlled horticultural environment. Of course, the level of automation varies significantly and for sure it cannot be taken as given in practice. In any case, and whatever any existing automation state, the lighting system is required to interact with the operation of control systems for irrigation, shading, relative humidity, heating and ventilation, and so on. Moreover, an additional advantage for the users of a lighting system is the provision of the lighting system provides some kind of remote control and monitoring. The feature of monitoring and control through specialized internet applications is very constructive and helpful.

6.2.2 Practical Requirements

Except of the functional requirements, there are also serious practical requirements for a lighting system in closed controlled horticultural environments. Even if all the functional requirements have been completely satisfied, ignoring practical issues may jeopardize the anticipated success of the production. Among the most significant practical requirements that have been identified are

(a) *Effective positioning*: according to the research studies on artificial lighting in horticulture, there are three types of lighting fixtures positioning methods, namely the *lighting from above* according to which the lighting fixtures are installed on the top of the cultivations, the *intra-canopy* (or else *inter-crop* or *inter-lighting*) according to which the lighting fixtures are installed within the canopy in various heights, and the *lighting from below* according to which the lighting fixtures are placed on the bottom, near the roots of the plants (Van Ieperen and Trouwborst 2008). These types of positioning ensure the maximum degree of freedom in order to overcome the well-known issues of shading and overheating. More specifically, as it has been reported, the intra-canopy lighting and lighting from below have the same or even better results in terms of the plants physiological properties and yield (Bergstrand et al. 2015; Tewolde et al. 2016; Van Ieperen and Trouwborst 2008). Hence, ideally, the lighting systems have to be light weight, as small as possible in their physical dimensions, and easily mountable. On the other hand, the lighting fixtures should produce the minimum radiant heat in order to avoid plants' overheating.

(b) *Scalability*: an artificial lighting system has to provide scalability in terms of adding extra lighting fixtures with the same or different light spectrum. This requirement is very critical especially in the experimental horticultural setups or in small-scale experimentations.

(c) *Human–machine interface*: while it is neglected during the lighting design phase, the human–machine interface (HMI) is very critical for the users in regular basis, because it allows a controlled and secure access of the system's functions and modes of operation.
(e) *Tolerance and robustness*: the lighting systems must be tolerant to any threats from their operational environment, such as chemical threats, heat stresses, humidity, and so on. Furthermore, the lighting systems have to be reliable during the entire cultivation periods. For this reason, lighting fixtures should allow for periodic maintenance and service with the minimum interruptions in their operation.

6.2.3 Electrical Requirements

Electrical energy is one of the most significant components of the total production cost in a professional horticultural facility. For this reason, lighting systems should be selected not just based on the light quality and intensity, but also by taking into consideration the energy consumption of the lighting fixtures. In particular, the electrical requirements for the lighting systems are:

(a) *Energy saving*: the energy demand for the operation of the artificial lighting system in a closed controlled environment has to be kept to the minimum. Unfortunately, traditional light sources suffer from increased energy consumption which in many cases is comparable to that consumed for the heating of the facility. For this reason, new and challenging technologies such as *geothermy* and renewable energy-based *micro-grids* (Kyriakarakos et al. 2015) have gained interest among greenhouse engineers as a response to the minimization of the total energy balance and the reduction of the electricity cost. The function of time programming of the lighting systems, mentioned above, can support the energy reduction in the sense of operating the artificial lighting scenarios during nighttime, where electricity rates are significantly lower in many countries (Tewolde et al. 2016). In general, a lighting system should have the highest energy efficiency possible. This means that the majority of the energy consumed by the lighting system should be transformed into light flux and not into unwanted heat, which, in certain cases could also burn the illuminated plants.
(b) *Cabling tidiness and safety*: the cabling of the lighting system has to be routed orderly so as to prevent any obstruction in the operations taking place daily on the spot. On the other hand, it is better to avoid high-voltage signals in the lighting systems' cabling, in order to increase the level of safety in the closed controlled environment. Especially in cases where the artificial lighting systems are used by amateurs in areas like indoor gardens, then extra caution with the cabling is required needed. In general, from the aspect of safety respect, it is better to have lighting fixtures that operate in relatively low DC voltages.

6.2.4 Economical Requirements

In regards to the economical requirements, there are several factors that have a direct or indirect influence on the cost of a lighting system, both in short- and in long-term time span (Tewolde et al. 2016). The most important requirements of a lighting system are

(a) *Low initial investment cost*: of course, this requirement can be satisfied in the early, pre-commissioning phase, with undertaking serious market solutions, analyses, and comparisons. Moreover, it is mandatory to avoid any kind of tradeoffs in the functions and the quality of the lighting system that could compromise the contemplated application.
(b) *Low cost of ownership*: except of the cost of electricity for operating a lighting system, there are also some implicit factors that can influence the cost of ownership of the artificial lighting applications. In particular, a lighting system should be serviceable without spending much money. Additionally, the ideal lighting system should allow for its spare parts and consumables to have second source distribution. Unfortunately, closed and turn-key solutions from sole sources impose significant costs in the long term. Hence, lighting systems should provide a sufficient degree of compatibility with other market solutions. This is particularly critical for the consumables, e.g., the light sources.

6.2.5 The Technological Context of Artificial Lighting Applications in Horticulture

According to the identified requirements presented above, it is clear that the operation of an artificial lighting system is not a trivial task. Even if there is sufficient expertise to design and deploy the right lighting system for certain cultivation, the expected results may significantly vary in practice, eventually. The reason for this lies on the fact that plants and their growth environment constitute a very complex, uncertain, and dynamically changing system, prone to climate changes. On the other hand, in the case of industrial greenhouses, there are also strong influences from market demand variations. Of course, the degree of such a contribution should be aligned with the return of investment. The full control of the plants' physiological and other properties is the aim of the urban horticulture for the twenty-first century. Towards this direction, it is mandatory for the lighting systems to exploit all the possible technological challenges and capabilities available today in order to make a step further to the artificial lighting, namely to move into the smart precision lighting (SPL). Future lighting systems will be able to seamlessly provide the right light in the right time and in the right place. The suggested roadmap to achieve SPL is depicted in Fig. 6.4. According to this suggestion, the first step includes the exploitation of any new challenging sources of light, the second step is associated with the engagement of networking and communication technologies in order to make the lighting

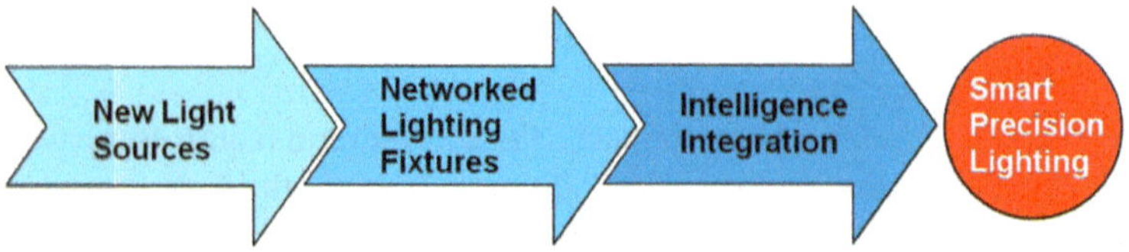

Fig. 6.4 The steps towards the achievement of smart precision lighting (SPL)

fixtures interconnected and remote controlled, whereas the third step includes the development and the integration of intelligence in order to realize the SPL devices.

6.3 Novel Lighting Sources

Typically, a lighting fixture, or else, a luminaire, is comprised of a lamp, a reflector, a power unit, and housing. There are several technologies of lighting sources that have been implemented in the form of various types of lamps during the last decades. Although any lamp is considered as a component that accepts electrical energy and produces illumination, there are certain characteristic differences among them, in regards to efficiency, operation longevity, intensity, quality of radiated spectrum, form factor, and cost (Kitsinelis and Kitsinelis 2015). Despite the fact that any given type of lamp can be met in various configurations in sake of improvement, however it is difficult to escape from the inherent disadvantages of the nature of its principles of operation. For the future horticulture, the integration of new lighting sources is very important, because this can improve the operation and performance of the artificial lighting applications. Taking into consideration the diverse requirements identified and presented in the previous section, modern horticulture should be relieved of the burden that insufficient types of lamps entail.

6.3.1 Traditional Lighting Solutions

Traditionally, the types of lamps that have been intensively used in horticulture for artificial lighting are the incandescent lamps, the fluorescent lamps, and the discharge lamps. The principle of operation of incandescent lamps is based on light generation from the heating of a specific metal filament of tungsten. From the electrical energy provided to an incandescent lamp only 15% is radiated as photosynthetically active radiation (PAR), whereas the rest 85% is emitted in the infrared spectrum above 850 nm. In terms of horticulture, this type of lamp is not suitable for photosynthesis due to the fact that it produces too much infrared radiation, so it cannot be considered as a sound source for supplemental lighting applications. On the other hand, incandescent lamps are used for night breaks to extend the growth day. Practically,

Table 6.1 Efficiency and lifetime of the most common types of lamps used in horticulture

Lamp	Efficiency (lm/W)	Working life (hours) [K]
Incandescent	12–26	1–3
Fluorescent (FL)	52–84	10–20
Metal halide (HID-MH)	80–90	10–20
High pressure sodium (HID-HPS)	117	10–30

incandescent lamps are low cost and easy to install, but at they suffer from short lifetime and energy efficiency. In addition, there are certain strategies to stop the production of incandescent lamps in sake of environmental preservation. Therefore, this particular type of light source cannot be considered as an option for the future horticulture.

The principle of operation of fluorescent lamps is based on the application of high voltage at the ends of a glass tube that contains mercury vapor in low pressure mixed with inert gases. The excitation of the enclosed elements by the high voltage produces light radiation. In this process, the glass of the tubes with fluorescent coverings plays a significant role. This type of lamp is very suitable for supplemental lighting in horticulture, because the light output has a fairly homogeneous distributed spectrum. Additionally, the cost for fluorescent lamps is quite low. However, fluorescent lamps suffer in the stability of their light output, which varies according to the temperature of the environment as well as according to the hours of operation. Specifically, fluorescent lamps may lose even up to 40% of their initial intensity after reaching 10,000 h of operation. As far as temperature influence is concerned, a fluorescent lamp operates fine when the environment of operation is about 35 °C, whereas it loses up to 50% in cases of temperatures below 16 °C. Therefore, fluorescent lamps appear to be unstable as a lighting source for smart precision lighting applications.

High-intensity discharge (HID) lamps are the most used type of lamps in horticulture today. Their principle of operation is considered as similar to that of fluorescent lamps, with the difference that the mixture of gases is in higher pressure. The most common HID lamps are the metal halide (MH) lamps and the high-pressure sodium lamps (HPS). MH lamps provide the most distributed spectrum but they are less efficient in energy conversion in PAR compared to the HPS lamps. In general, HID lamps are very suitable for supplemental as well as replacement lighting applications.

All of the aforementioned types of lamps impose limitations to the horticultural lighting applications. In particular, they produce a significant amount of heat, resulting in the increase of the area's temperature. Incandescent and HID lamps in particular, can burn the cultivated plants if they are installed in a close distance with them. In addition, due to their bulky form factor, they create significant shadowing effects to the plants. On the other hand, they lag in terms of efficiency in the conversion of electrical energy to light flux, as it is shown in Table 6.1. Therefore, the traditional lighting sources, although they are low cost, they do not support the requirements of horticulture for the smart, precise, and energy-efficient lighting.

6.3.2 Technological Challenges in Lighting Sources

The most significant technological challenge in lighting is the use of the light emitting diodes (LEDs) as lighting sources. LEDs produce light when a certain amount of direct current is flowing from anode to cathode electrodes. These devices have been vastly used as screen components in various electronic devices for many years, but only during the last decade their design has been improved to allow their usage as a main source for residential and industrial lighting applications (Kitsinelis and Kitsinelis 2015; Walerczyk 2014). Today, lighting that is based on LEDs is known as solid-state lighting (SSL), due to the fact that the light source is a solid-state electronic component, in contrast with the traditional lighting sources.

6.3.2.1 Prospects and Expectations of SSL in Horticulture

From its infancy, SSL triggered enormous financial and research resources in order to grasp the opportunity of the upcoming lighting revolution. Initially, SSL applications focused on residential (e.g., architectural lighting, households' lighting, etc.) and industrial (e.g., entertainment, automotive, etc.). In horticulture, SSL started to be used with a significant delay. Even today, HPS lamps are still the most commonly used light source in commercial horticulture (Van Ieperen and Trouwborst 2008). The reason behind this is probably that professional horticulturists are skeptical and conservative in terms of investing to new technologies, and they actually prefer to buy and install ready-tested and almost mature solutions. During the last decade, LEDs have improved their characteristics associated with their performance and operation cost. Therefore, existing LED-based SSL solutions contradict those who were skeptical on whether LEDs could surpass HPS lamps (Bergstrand et al. 2015; Singh et al. 2015).

6.3.2.2 LEDs Technology Advantages Over Existing Lighting Sources

Lighting fixtures that are based on LED technology are by far more energy efficient compared with all the existing lighting solutions (Singh et al. 2015). Specifically, LED technology may ensure up to 80% lower energy consumption for the same light intensity. Therefore, for horticultural applications where the electricity cost is almost 25% of the overall production cost, such achievements are very important. In terms of light intensity, existing LEDs can reach up to 190 lm/W (CREE 2017), whereas in the near future they will reach 300 lm/W. Therefore, this technology can successfully support supplemental lighting applications in horticulture. As a proof of this, in (CREE 2016) there is a reference design of a Haybay LED-based fixture which surpasses the light intensity of an HPS lamp-based fixture. On the other hand, in regards with light quality of LEDs, in contrast with their competitive traditional light sources, they produce light in a relatively narrow wavelength range, and their light flux is of higher relative quantum efficiency (RQE). Moreover, being solid-state components, LEDs allow for the maximum controllability in terms of their spectral composition. So far, the benefits of LED-based SSL in plants growth have been studied by several researchers with significantly positive results (Massa et al. 2008; Singh et al. 2015). On top of the aforementioned advantages of LEDs' technology, is their longevity which, in the long term can compensate their cost (Singh et al. 2015).

6.3.2.3 LEDs Technology Advantages for Closed Controlled Horticultural Environments

As it is reported by many (Massa et al. 2008; Morrow 2008), LED lighting can variously benefit controlled horticulture. Significant gain can be derived from the fact that LEDs do not generate radiant heat that could harm plants' foliage. In particular, LEDs infrared intensity is a very controllable parameter, in contrast with other existing lamp technologies. Also, LEDs operate in lower temperatures (around 80–90 °C) in contrast with incandescent lamps (2500 °C) and the various types of HID lamps (200 °C). Consequently, LED-based fixtures fulfill the flexibility and versatility required in intra-canopy and bottom lighting applications in closed controlled horticultural environments. In addition, from the deployment aspect, LED-based lighting support the reduction of high-powerhigh-power cabling, leading to cables' cost reduction and safer wires' routing across the closed horticultural environments.

6.3.3 Existing LED-Based Applications in Horticulture Today

In recent years, LED-based artificial lighting has been significantly studied by many researchers in horticulture as well in floricultural applications (Singh et al. 2015). In summary, the main areas of such studies fall into three application areas, namely

(a) *usage of SSL in combination with traditional lamps* for supplemental lighting (Wojciechowska et al. 2015);
(b) *LEDs light effects on plants* (Jeong et al. 2014; Meng and Runkle 2015; Samuoliene et al. 2015; Piovene et al. 2015; Taulavuori et al. 2016); and
(c) *LEDs light effect on transplants* (Brazaityte et al. 2009; Hernâdez and Kubota 2016).

In the majority of the existing research studies worldwide it is obvious that the ability of LED lighting to control the light spectrum allow the control of various important quality and quantity parameters of plants. On the other hand, all studies emphasize the superiority of LED lighting regarding th energy reduction especially compared with traditional types of lamps.

In recent years, LED-based artificial lighting has been significantly studied by many researchers in horticulture as well in floricultural applications (Singh et al.

2015). In summary, the main areas of such studies fall into three application areas, namely the usage of SSL in combination with traditional lamps for supplemental lighting, LEDs light effects of plants, and LEDs light effect on transplants. The emphasis of research in the majority of works was on the light wavelength optimization. The most representative studies are given below.

Wojciechowska et al. (2015) evaluated the effects of LED and HPS lights as a supplemental to solar radiation in winter growing of greenhouse lamb's lettuce plants. More specifically, sole red, combination of red and blue in several different ratios, and warm white LEDs were investigated, together with HPS lamps. The results showed that at harvest stage, fresh weight of rosettes, dry matter, soluble sugars, total phenols content, and radical scavenging activity were the greatest under the 90/10 Red/Blue ratio of LED lamps, while the least efficient in the enhancing of lamb's lettuce yield proved to be the supplemental lighting with white LED lamps. Jeong et al. (2014) examined the effects of blue spectrum supplemental lighting on flower bud formation and stem elongation growth of cut chrysanthemum plants cultivated in greenhouses. According to their findings based on several LED light treatments, supplemental blue light, at least in part, may promote stem and internode elongation growth without any inhibitory effect on flower bud formation. Meng and Runkle (2015) assessed the effect of spectral quality of photoperiodic lighting on greenhouse plants, by using LEDs of different spectral ranges at different periods during the day. Their extended experiments on the flowering of five short-day plants and two long-day plants showed that low-intensity blue light (400–500 nm) during the night did not influence flowering, while white LEDs that emit little far-red light (700–800 nm) were effective at creating long days for short-day plants but only for some long-day plants. Samuoliene et al. (2015) evaluated the effect of different combinations of red, blue and far-red LEDs on the variation of photosynthetic pigments, non-structural carbohydrates and flowering initiation of radish, carrot and strawberry plants. They concluded that spectral quality of light influenced the morphogenesis and diverse physiological responses of all investigated plants, while, in each case, flower initiation processes could be controlled by tailoring the illumination spectrum. Piovene et al. (2015) investigated the physiological and phytochemical plant responses to LED lights in indoor cultivation of leafy and fruit vegetable crops (sweet basil and strawberry), with the final aim of improving both productivity and nutritional quality. In all experiments, associated LED treatments led to increased plant biomass, fruit yield and energy use efficiency, compared to the traditional fluorescent lamps. The authors concluded that a spectral red:blue ratio of 0.7 was necessary for proper plant development and improved nutraceutical properties in both crops. Taulavuori et al. (2016) examined species-specific differences in synthesis of flavonoids and phenolic acids under lengthening periods of enhanced blue light in greenhouse cultivation of red leaf lettuce and basil plants, with an aim to compare red and blue weighted light spectra in relation to biosynthesis of the compounds. Flavonoid and phenolic acid biosynthesis in plants were found to be species dependent. Red leaf lettuce was much more responsive to supplemental blue light. Based on these results, the authors suggested that both blue and red light may be needed to regulate the accumulation of phenolics in basil. Finally, some of the compounds detected accumulated continuously as a function of the spent time under supplemental blue light in red leaf lettuce, but not in basil. Brazaityte et al. (2009) studied the effect of various combinations of LED lighting spectra on the growth-chamber cultivation of cucumber transplants. The comparison with the use of HPS lamps showed that supplemental lighting with green or orange LEDs could accelerate the growth of the transplants till technical maturity, and decrease the required energy input. Hernández and Kubota (2016) evaluated cucumber seedlings physiological responses to different blue and red photon flux ratios using LEDs. Results showed that hypocotyl length decreased with the increase of blue light or red light, up to a blue/red ration of 75/25. However, plants under solely blue light had unexpected greater plant height, hypocotyl, and epicotyl length, and leaf area than plants under all other treatments. Chlorophyll content per leaf area, net photosynthetic rate, and stomatal conductance increased with the increase of blue light, while shoot dry and fresh mass decreased with the increase of blue when plants were irradiated with the combination of blue and red light. The addition of green light to the spectrum did not have any influence in cucumber plant responses. The authors concluded that more research is needed to find the optimal spectrum for the growth and development of horticultural crops under sole source electrical lighting such as LEDs.

6.3.4 Design Advantages for the Smart Precision Lighting in Horticulture

Except of the advantages of LEDs technology over the traditional light sources, which are basically based on their principle of operation, it also appears to be a very convenient technology in terms of the light fixtures design. Design options such as the inherent light spectrum composition control, freedom in form factor and physical dimensions, scalability in light intensity, and the configuration and definition of the light flux angles, which are natively supported by LEDs technology, can help the designers to focus on the particular needs of closed and controlled horticultural environments, and create the most optimized light fixtures for the smart precision lighting.

6.3.4.1 Design Options in Spectrum Configuration

Traditionally, in order to configure the spectrum of the artificial light, different light fixtures with monochromatic lamps are combined in various ratios. Of course, this option is supported by LED-based light fixtures. Moreover, LEDs technology allows the design of multi-color fixtures. In this case, designers can mix LEDs of different light wavelengths in the same lamp. Alternatively, multi-core and multi-color LEDs can be employed. Such an LED is illustrated in Fig. 6.5. Therefore, SSL based on

Fig. 6.5 The quad-core LED CREE MC-E Color with four distinct LEDs (red, green, blue, and white) with luminous efficacy of 99 lm/W and maximum power of 9.5 W

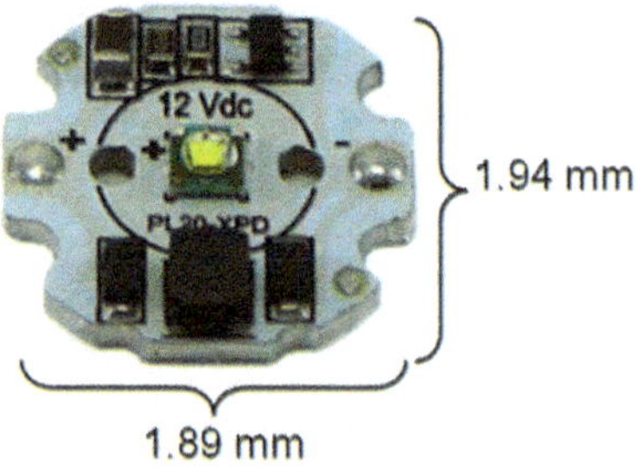

Fig. 6.6 Small-sized-high-power (4 W) single-LED (CREE XP-G) aluminum board lamp of 275 lm luminous flux and very small form factor

LEDs, supports the design of optimized spectrum solutions, and it eliminates the need for procurement of many different light fixtures for the same application.

6.3.4.2 Design Options in Form Factor and Physical Dimensions

As it has been pointed out above, there are several shading and overheating issues caused by the traditional bulky and hot light fixtures. Ideally, a light fixture should radiate the minimum heat and its physical dimensions should be the optimum in order to avoid shading effects on the illuminated plants (Bergstrand et al. 2015). In their turn, LEDs are handled as plain surface-mount devices which can be soldered onto a printed-circuit board (PCB). This capability allows designers to develop light fixtures with optimized shape, dimensions to satisfy the requirements of positioning and effectiveness in various urban horticulture real-life applications. In addition, the usage of aluminum PCB material instead of the commonly used FR4 glass-reinforced epoxy laminate material helps the heat produced by LEDs to be directly dissipated and transferred to smaller heat sinks. Given any particular requirements for light intensity, few high-power LEDs can be used in smaller area PCBs, or more mid-power LEDs can be used in larger area PCBs or in. Figure 6.6 illustrates the implementation of miniaturized single high-power LED light engine using aluminum PCB. Figure 6.7 shows the PCB design of a circular LED engine with 12 high-power LEDs for supplemental lighting applications. Both design examples were developed in the Piraeus University of Applied Sciences (PUAS) laboratories.

Fig. 6.7 The PCB design of a LED engine with twelve high-powerhigh-power LEDs in circular shape of 90 mm diameter

Fig. 6.8 A typical high-powerhigh-power white color LED (CREE XP-G) with luminous efficacy of 146 lm/W and maximum power of 5 W

6.3.4.3 Design Options in Light Intensity

Similarly to all diodes, LEDs have to be biased with a certain forward voltage in order to allow the flow of a direct current through their body. Specifically, in the case of LEDs the forward voltage ranges up to 3.5 V DC, whereas the direct current can vary from few milliamperes up to 2 A. Unfortunately, the higher power does not mean better energy efficiency in most cases. The selection of the specific LED type depends on the particular application. The improvement of the luminous efficacy, which is the ratio of Lumens per Watt (lm/W) is in high priority in the agenda of semiconductors industry worldwide. Today, LED lighting fixtures designers can make their choices from a vast range of commercially available LEDs with various light intensity, various energy efficiency, and various physical dimensions in order to cover the existing but also the future requirements of artificial lighting in closed and controlled horticultural environments. Figure 6.8 shows a representative high-powerhigh-power white color LED, whereas a mid-power white color LED is presented in Fig. 6.9.

Except of the aforementioned examples of white color LEDs, there are also monochromatic LEDs namely, blue, royal blue, green, orange, red, far-red, and

Fig. 6.9 Typical mid-power white color LED (OSRAM Duris E3) with luminous efficacy of 153 lm/W and maximum power of 0.5 W

Fig. 6.10 Typical COB ultra-high-power white color LED (SHARP GW7GAL50SGC) with luminous efficacy of 143 lm/W, maximum power of 25 W, and maximum light of 14,000 lm

infrared LEDs in various light intensities and power. Additionally, when there is the need for extra high levels of power, then the alternative design option of using the so called chip-on-board (COB) LED solutions is followed. More specifically, COBs are several LEDs connected in various combinations together, soldered onto the same PCB and optically matched under the same primary lens. Such a COB LED solution is presented in Fig. 6.10.

6.3.4.4 Design Options in Light Flux Angle

Traditional lighting solutions impose their characteristics in the luminous flux angles. These constraints together with the radiation of high temperatures from lighting fixtures to their environment cause serious restrictions in the positioning of fixtures in the horticultural environments. Therefore, there are many tradeoffs in deployments in terms of the optimum positions of the fixtures. LEDs solutions are flexible and convenient because they allow the control of their viewing angles through the use of low cost secondary optics. In particular, specific plastic lenses (see Figs. 6.11 and 6.12) can be used on top of the LED components within the lighting fixture which deterministically define the viewing angle of the produced light flux.

With the aforementioned design options it is easier than ever to control the viewing angle according to the preferred height of distance of placing the light fixtures. Therefore, LEDs fully support the control of the critical factor of lighting geometry (Massa et al. 2008) within closed and controlled horticultural environments.

Fig. 6.11 Typical plastic lenses for single-LED use (courtesy of Ledil)

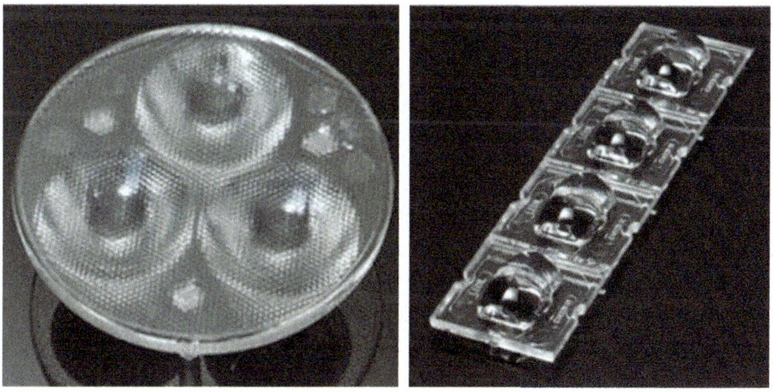

Fig. 6.12 Typical multi-LED plastic lenses (courtesy of Ledil)

6.3.4.5 Solid-State Lighting Design Cases

In order to demonstrate the pluralism in the design options of LED-based lamps, or else LED engines, two indicative design cases were made in PUAS, one with several mid-power LEDs and the other with few high-power LEDs. Both LED engines can be used either for white or other specific color. The LED engine of mid-power can be an alternative to fluorescent lamps as long as they are cascaded and connected together. Regarding the secondary optics for the viewing angle configuration, in the design case A (mid-power) a single diffuser implemented on the plastic or glass front cover of the fixture is used, whereas in the design case B (high-power) each LED can accept its own lens part. The two design cases are illustrated in Fig. 6.13. Table 6.2 quotes the technical characteristics for the two contemplated LED engines, while Figs. 6.14 and 6.15 present the particular measurements and calculations made for each one of the designed LED engines.

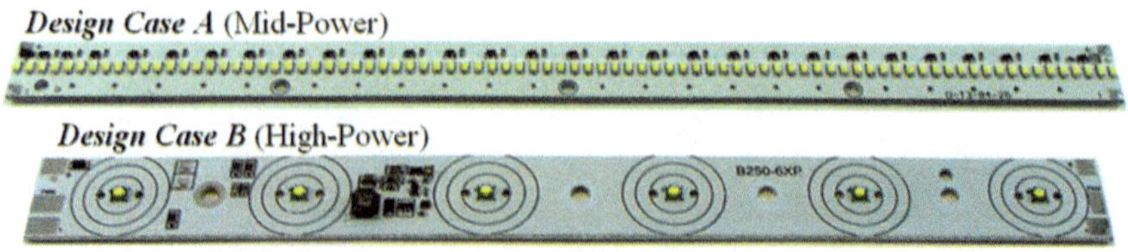

Fig. 6.13 Two different implementations of LED engines

Table 6.2 Technical specifications of two representative LED engines, one using many mid-power LEDs and another using few high-power LEDs

Specifications	Design case A	Design case B
LED type	OSRAM Duris E3	CREE XP-G
Supply voltage	12 Vdc	20–24 Vdc
Power	5.6 W	22 W
Power class	Mid-power	High-power
PCB material	FR4	Aluminum
Dimensions	250 mm × 16 mm	230 mm × 20 mm
Number of LEDs	84	6
Luminous Flux @ 6000 K	530 lm	1650 lm
Luminous Flux @ 5000 K	508 lm	1440 lm
Luminous Flux @ 3000 K	493 lm	1180 lm

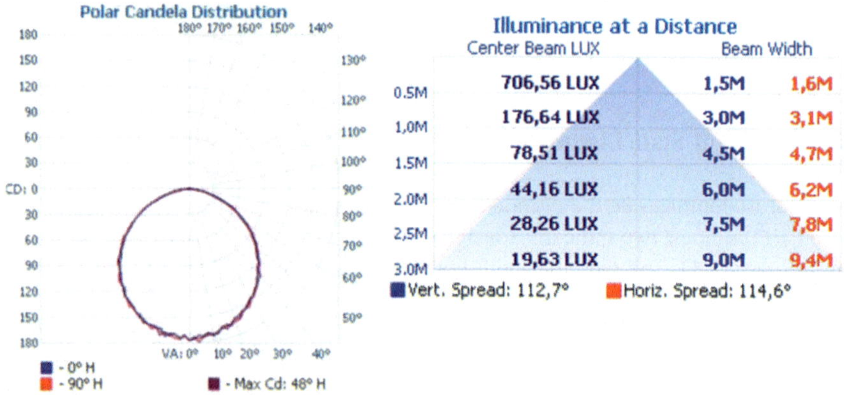

Fig. 6.14 Measurements and calculations for light intensity and geometry of the mid-power LED engine at 6000 K color temperature

6.4 Networked Lighting Fixtures

The networking function of lighting fixtures is very important for the artificial lighting applications in various closed and controlled horticultural environments, because

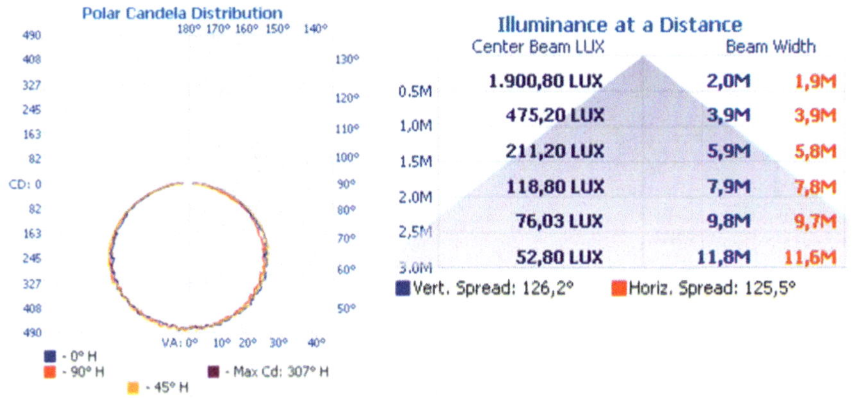

Fig. 6.15 Measurements and calculations for light intensity and geometry of the high-power LED engine at 6000 K color temperature

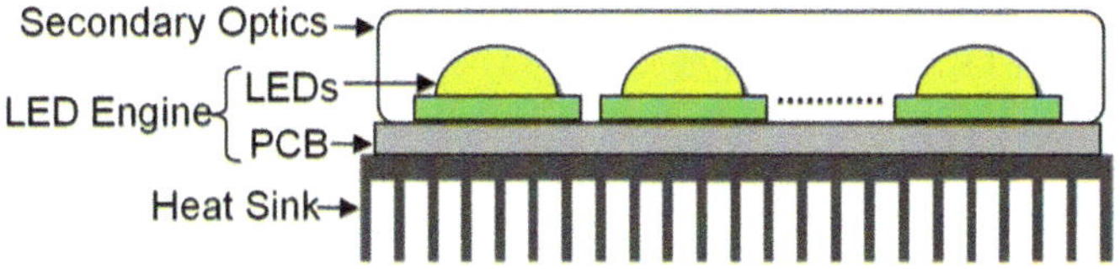

Fig. 6.16 The simplest structure of a LED lighting fixture (LF)

it ensures that every single fixture is addressed and commanded by a central controller which is executing the overall lighting scenario. Reasonably, the function of networking is closely related to the function of the energy management (Piromalis et al. 2016). Fortunately, there are alternatives in terms of design options in lighting systems implementation (DiLouie 2006). However, such decisions may entail tradeoffs and compromises in the ultimate lighting implementations.

6.4.1 Energy Management Approaches

The simplest implementation of a LED lighting fixture (LLF), as is shown in Fig. 6.16, consists of three main parts: the LED engine; the secondary optics; and the heat sink for releasing the PCB of the LED engine from LEDs operation temperatures. Notably, in this case there is no power supply unit in the LLF. In this case, LEDs are connected each other in various combinations of serial and parallel connections and an external power supply unit with constant current output capability is required in order to make the LED engine to produce the preferred illumination.

Practically, all LED engines require constant current so as to ensure a homogeneous light output from all of their LED devices. The way that this constant current

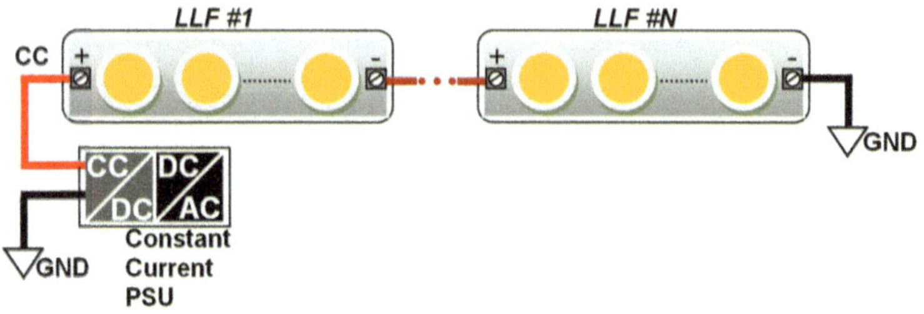

Fig. 6.17 Block diagram of the constant current (CC) power distribution scheme

is provided to the LLFs determines the power distribution topology of the lighting application. Specifically, there are three main power distribution schemes, namely the constant current (CC), the retrofit, and the constant voltage (CV) scheme. Each one of these has its own advantages and disadvantages for the lighting applications in horticulture.

According to the CC scheme, the LLFs can be similar with the simplest implementation shown in Fig. 6.16. This topology allows the LLFs to be easily cascaded, while it supports the wiring order (see Fig. 6.17). Unfortunately, CC power supply units are expensive and very often they impose limitations in the number of the cascaded LLFs. Moreover, it is not possible for connecting LLFs from different manufacturers due to the fact that each LLF requires different values of constant current. In other words, there are expandability and maintenance constraints for the applications in the long term.

Alternatively, the retrofit scheme simplifies the collaboration of unlimited number of different LLFs (see Fig. 6.18). In this particular case, the LLFs accept power from the AC mains and therefore they must include in their enclosure all the necessary power conversion circuits. Although the retrofit scheme appears to be very convenient to horticultural domain of lighting applications, especially because the traditional lamps are all AC powered, there are issues to be considered. Specifically, the cost of the LLF is reasonably higher than a LF which contains just the LED engine, and the guaranteed LEDs longevity very often is compromised by the working life of the capacitors and other components of the integrated power supplies. Hence, retrofit scheme can jeopardize the cost of investment in the long term.

Regarding the third power distribution scheme, that of constant voltage (CV), the difference is that the LLFs integrate just the circuits for the conversion of a common DC voltage to the constant current that is needed for the particular hosted LED engine (see Fig. 6.19) (Thomas et al. 2012). The most important advantage of this topology is that common, inexpensive, constant voltage external power supply units can be employed. As it regards with the application deployment, there the CV scheme requires the usage of DC cabling infrastructure.

From the initial investment point of view, as a rule of thumb, it is suggested that for small-scale applications, e.g., applications with few tens of LLFs, the CC power

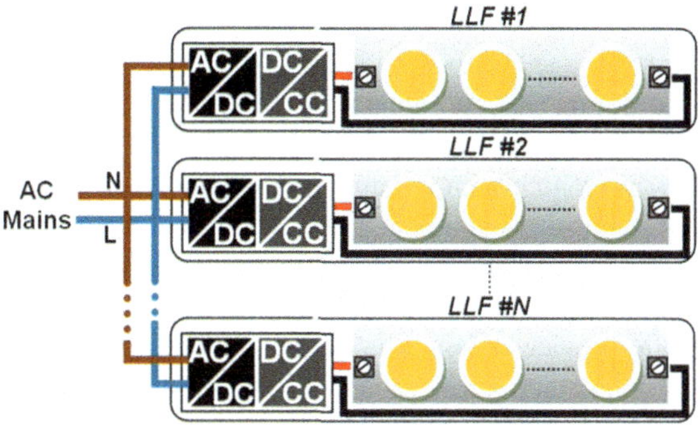

Fig. 6.18 Block diagram of the retrofit power distribution scheme

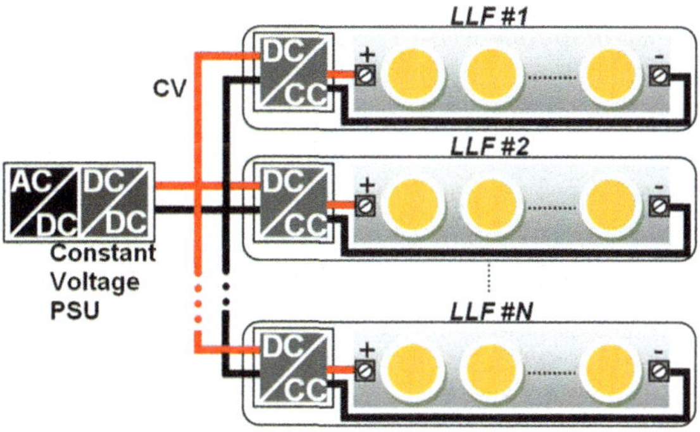

Fig. 6.19 Block diagram of the constant voltage (CV) power distribution scheme

distribution scheme is the most proper choice. Accordingly, for medium- and large-scale applications, such as those of professional horticulture, it is more cost effective to follow the CV scheme. Roughly, the cost save can be up to 20%.

Moreover, Piromalis et al. (2016) have suggested a new architecture for LED engines design, the so called LDoB (LED Driver on Board), according to which the DC voltage to constant current circuitry is implemented onto the very same PCB of the LED engine. Amongst the main advantages of the LDoB approach are: the self-protection of the LLF against over-heating and power problems such as broken, as well as, short circuits; the support for multi-color LED engines power management; and the ability to decrease the form factor of the LLFs. The power distribution of LDoB-based LLFs is shown in Fig. 6.20.

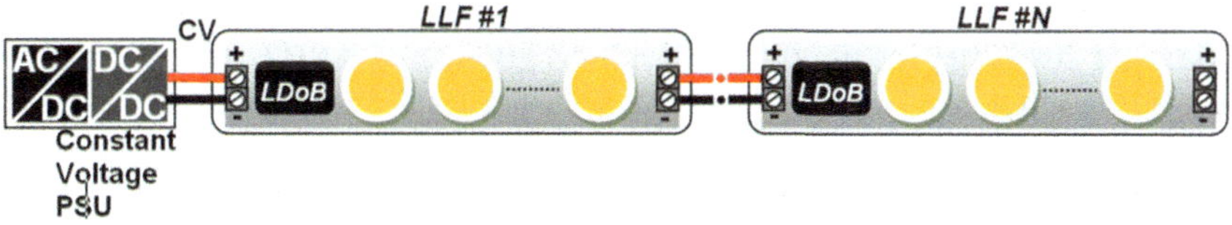

Fig. 6.20 Block diagram of the power distribution to a string of LDoB-based LLFs

6.4.2 Technologies for Interconnected Lighting Devices

The notion of interconnected LLFs has its roots in building management systems (BMS) where the full control of every single LLF is at the service of living and working environment improvement. As solid-state devices, LEDs permit for unparalleled controllability (Simpson 2003) which can support the research for applications such as the intelligent ambient living and so on (Huynh et al. 2011). On the other hand, until recently, horticultural lighting systems were considered as static on–off controlled devices. Hopefully, for economic reasons, the operation timing and dimming of illuminance were introduced in order to reduce the cost of electricity. Of course, the challenges of exploiting the dynamically controlled LEDs for smart precision lighting applications in horticulture opens new horizon for every kind of professional or amateur applications.

Regarding the technological solutions for LLFs interconnection, i.e., networking, there are several longstanding alternatives mainly based on networking buses of standardized communication protocols, such as the EIB (European Installation Bus), or LON (Local Operation Network), and KNX (an open standard EN 50090, ISO/IEC 14543) (Association, LonMark Interoperability 2002; Cho and Dhingra 2008). Additionally, other simpler protocols have been introduced for networked LLFs, such as the DALI (Digital Addressable Lighting Interface) and DMX512 (IES, IESNA 2011). In general, an interconnected LLF must integrate a microcontroller-based circuitry which is responsible for the physical interface with the lighting network and also determines various data commands sent from the application's master control device to operation parameters of the LLF. Regarding the physical layer of the networking, the majority of the aforementioned protocols use the standard bus of EIA-485 (López et al. 2007). This is a four-wire bus (power, ground and two differential signals for addressing and data) which can interconnect up to 32 devices. In cases where there are many devices then the EIA-485 bus can be extended through the use of special extenders (TI 2002). Alternatively, for lighting applications with increased complexity and time-critical dynamic response in terms of the networked LLFs, DMX515 and other similar protocols use for instance Ethernet extensions (Newton 2005). In general, the wired buses impose the usage of specific multi-core cabling and they negatively affect the cost of ownership and deployment of applications. As it is illustrated in Fig. 6.21, the interconnection of a LLF requires up to 10 distinct cables.

Almost the last decade there is an explosion of the wireless connectivity. Therefore wired buses can be replaced by communication protocols either coming from the IT domain such as the WiFi, or from the domain of the wireless sensors networks (WSN) (Maxfield 2011) such as the IEEE 802.15.4/ZigBee, the DASH7, the wireless Hart and wireless M-bus (Piromalis et al. 2013; Piromalis and Arvanitis 2016b; Yurish and Cañete 2013). Respectively, the structure of Fig. 6.21 has to be modified in terms of replacing the wire bus interface circuitry with a radio frequency (RF) transceiver in order to achieve the wireless networking of the LLF.

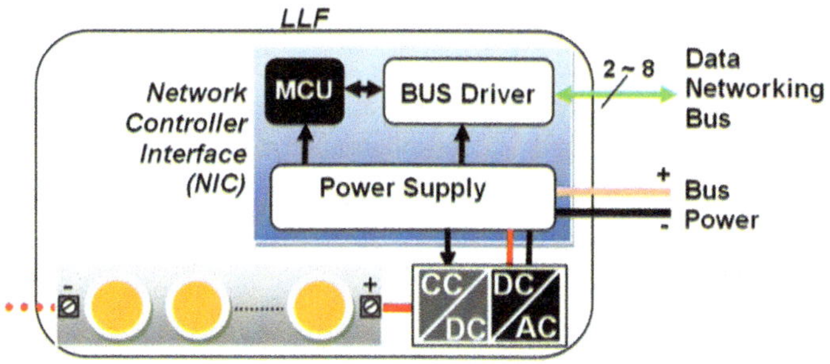

Fig. 6.21 Generic structure of a wired networked LLF

Although wireless connectivity is more convenient regarding the cabling and that they allow for two-way communication amongst the interconnected LLFs, they are orders of magnitude more expensive than their counterparts wired interfaces. For horticulture where the number of LLFs may be high, the approach of autonomous interconnected lighting devices is under questions due to the fact that there is no need for lighting variation in fine resolution as it is required in home or office applications.

6.4.3 Current Commercial Solutions for Artificial Lighting in Horticulture

Current solid-state lighting solutions for horticulture are dominated by the well-established lamps manufacturers mainly because they used to be the sole vendors of the traditional lamps for many decades. The proposed lighting systems are following the constant current power distribution scheme within which they wired or wireless connectivity interfaces are integrated. The advantage of this approach is that LLFs do not have to include their own networking interfaces, and so they can be more competitive in terms of cost. On the other hand, this approach suffers from the limitations and constraints presented in Sect. 6.3 for CC scheme. Moreover, commercial solutions have the following characteristics:

- they are single-source, turn-key solutions of closed architectures,
- they provide limited resources to third-party developers,
- they impose practical barriers in terms of combining LLFs of other vendors,
- they tend to employ proprietary protocols instead of standardized protocols for the networking of the LLFs,
- they are rather rigid regarding the expansion of the LLFs number or capabilities,
- they impose limitations towards the collaboration of the lighting system to a central automation system,
- they do not allow for the integration of any new technology after deployment, and
- they do not support the particular requirements of the horticultural research.

Looking into the future, it is mandatory to have commercial solutions which are based on open-source resources in order to be accessible from the various stakeholders towards optimized and effective lighting applications in horticulture. Versatility, flexibility, reconfigurability, and scalability should be the priorities of the designers of tomorrow.

6.4.4 Towards the Networked Clusters of Lighting Devices

Basically, the grouping of LKFs in clusters is very important for the cost reduction of the lighting application and also for the support of the maintenance of the LLFs and their network (Smet et al. 2012). In other words, it is more convenient to network and control several cluster-head devices than many autonomous LLFs. In practice, the commercially available lighting solutions are targeting to meet this requirement by providing controllable power supply units (CPSUs) capable of supplying several LLFs simultaneously. Professional horticulture is very familiar with the concept of clusters because the plants are organized in well-defined zones of rows. Therefore, lighting systems should be installed according to the clusters of plants. Following these approach farmers could control the artificial illumination occasionally. However, the almost established approach of CPSUs cannot allow for the granularity of control required in small and custom-made urban horticulture applications, as well as for research horticultural setups.

Towards the applications of smart precision lighting for the future horticulture it is proposed the exploitation of hybrid wired and wireless networks. In particular, the cluster-heads should be wirelessly networked whereas the cluster member LLFs should be interconnected with their cluster-head through simple and low-cost wired networking buses. Therefore, cluster-heads should be designed as data gateways. Piromalis et al. (2016) have proposed a network architecture for the smart precision lighting in horticulture based on the single-wire LIN-bus (LIN, Consortium 2006) for the interconnection of the cluster member LLFs and for any of the existing standardized protocols such as WiFi or IEEE 802.15.4/ZigBee for the wireless networking of the cluster-heads. Figure 6.22 illustrates the block diagram of the LEDWIRE networking architecture. The LWG and LWS in Fig. 6.22 stand for the LEDWIRE Gateway and LEDWIRE Slave devices accordingly.

The advantages of such networking approach are

- the cabling networking and power distribution cabling complexity is eliminated,
- two-way communication among the LLFs of all the co-existed lighting clusters is ensured in order to not just broadcasting intensity and color commands to LLFs but also to read LLFs feedback for their individual state and health,
- supports smart precision lighting applications by achieving the highest granularity and control of LLFs, and
- easily designed in LED engines and it is very low cost to implement.

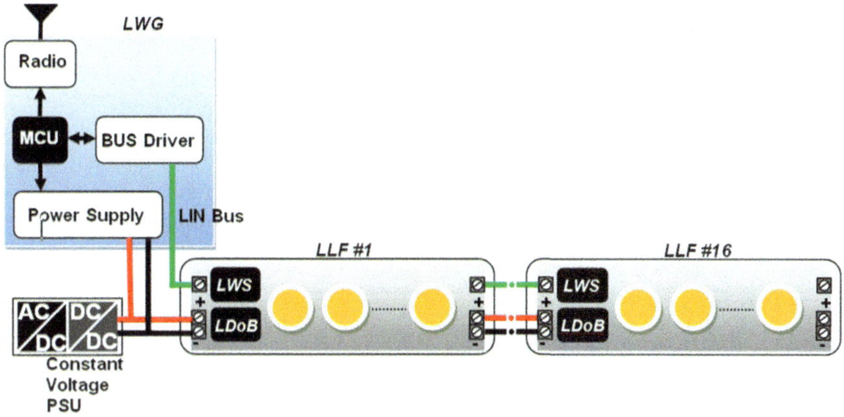

Fig. 6.22 Block diagram of the LEDWIRE networking architecture proposed for supporting smart precision lighting applications in horticulture

Table 6.3 Capabilites of changing the electrical and physical attributes of a cluster member LLF according to the CPSU and LEWIRE networking architectures

	Cluster member LLF control	
Attribute	CPSU	LEDWIRE
Luminance	No	Yes
Color	No	Yes
Color temperature	Yes	Yes
Time response	No	Yes
On/off state	No	Yes
Secondary optics	Yes	Yes
Physical dimensions	Yes	Yes
Energy consumption	No	Yes
Common power supply unit	Yes	Yes

Table 6.3 quotes the capabilities of LEDWIRE architecture against the existing controlled power supply units (CPSU) architecture in terms of the ability of changing the electrical and physical properties of a cluster member LLF. Obviously, networking architectures similar to LEDWIRE increase flexibility and controllability to the maximum degree.

6.5 The Integration of Intelligence Towards Smart Precision Lighting

The integration of intelligence into artificial lighting systems for horticulture presupposes three fundamental agents, namely sensors, models, and interaction with the automation system. At the same time, it is crucial that any effort towards smart precision lighting applications is based on standardization.

6.5.1 *Sensors for Smart Precision Lighting*

In order to take the right decisions on certain crop's requirements for light, it is necessary to be able to get measurements of the actual quality and quantity of light in situ. Thus, sensors are required in order to determine how much light and of what spectrum is need by the plants. Ideally, these measurements should be available in real time since natural light is changing dynamically in most cases. Traditionally, there are three main categories of lighting sensors that is *photometric sensors*, *pyranometers*, and *photosynthetically active radiation (PAR) sensors*. The contribution of each kind of sensors to horticulture varies.

In particular, photometric sensors have been designed basically for human vision applications, their units are usually in lux, lumens, and foot-candles, and they are used to measure the level of light intensity in a certain area. Such sensors, although they are relatively inexpensive and easy to use, they are only used for providing lighting systems with an indication regarding the intensity and uniformity of lighting on plants.

Pyranometers exist in various types to support different precision as well as deployment requirements (e.g., flat plane cosine-corrected pyranometers, flat measuring surface pyranometers, etc.). Regardless their type, pyranometers are used to measure the solar energy radiation and sky radiation. The measurements are in watts per square meter per second (W m^{-2} s^{-1}) and can be used to give an indication of the relative strength of solar radiation. In general, pyranometers are expensive and are usually installed outdoors.

PAR sensors, also known as quantum PAR or quantum photon sensors, measure the photosynthetically active radiation at certain points of the area (e.g., on top, on at the bottom or within the crop canopy). Specifically, these sensors measure the photon density contained in all the wavelengths of the spectral of 400–700 nm. The PAR sensors give their measurements in micromoles per square meter per second (μmol m^{-2} s^{-1}). PAR sensors are very useful in horticulture, especially for research purposes and in the design and planning of new or reconfiguration of existing artificial lighting applications. However, it is very difficult to distinguish between the various wavelengths of the spectrum contribution in PAR values. Therefore, shorter spectral bands allow for more precise PAR readouts. Similarly, in the cases of natural or supplemental lighting, where the wavelengths are of full spectrum, the PAR readouts can be used as a relative indication of the photon flux due to the fact that the entire spectrum is equally weighted.

Even though the conversion from light intensity, radiation, and quantum photon sensors is quite difficult and complex, it is suggested that a lighting system for smart precision lighting should employ various types of lighting sensors, to compensate potential inaccuracies entailed from their particularities. On the other hand, regarding sensors positioning, extra care should be taken because this can influence the accuracy of sensors validity. For instance, certain points in the installation may suffer from occasional shading or overheating. Both et al. (2002) have studied the implementation of three-dimension lighting maps in order to increase the level of information for the illuminated plants' environment. Maybe in the near future, specific autonomous mechatronic systems equipped with lighting sensors will be employed to acquire lighting measurements from the entire area of cultivation. Moreover, the evolution in the manufacturing of semiconductors for photo-spectroscopy can have a positive contribution to smart precision lighting, in the sense that new, miniaturized and lightweight light sensors will enhance the validity and integrity of lighting measurements.

6.5.2 Controls in Artificial Lighting Applications in Horticultural Domain

Light intensity inside a greenhouse, and more precisely, photosynthetic active radiation (PAR), is one of the most important parameters (together with temperature and relative humidity) that need to be controlled, as it drastically affects plant growth. However, unlike temperature and relative humidity which are usually precisely controlled by modern greenhouse environment management and control systems, light is rarely controlled in similar degree of precision, while daily light integral, typically, is never controlled (Albright et al. 2000). The provided light intensity is usually measured in daily light integral. Based on that metric, Albright et al. (2000) have developed an advanced algorithm for PAR control in greenhouses using prediction and real measurements acquired in specific time periods during the day. Also, Ferentinos and Albright (2005) have created a sophisticated computational methodology for the best fit design of lighting control system. Additionally, there are certain studies that investigate the design of lighting control taking into consideration the indulgence of the CO_2 generation within cultivations (Fierro et al. 1994; Both et al. 1998; Ferentinos et al. 2000; Ayari et al. 2000).

More recently, with the introduction of LEDs in horticultural supplemental lighting systems and their combination with traditional high-intensity discharge (HID) amps, lighting control is no longer solely focused on the quantity of light exposure (i.e., light intensity and integral), but also on light quality, which mainly concerns the provision of light at specific spectral regions. Furthermore, the combination of LED lighting with electronics and microelectronics, provides the means for more accurate and complex light intensity and integral control (Pinho et al. 2006). Several dynamic control methodologies have been proposed (Gómez and Mitchell 2014; Pinho et al. 2013; Van Iersel et al. 2016), leading to the development of sophisticated supplemental lighting control systems. These systems, in addition to the provision of precise quantities of light of specific quality according to greenhouse plants' needs, they can also achieve significant energy savings (Harbick et al. 2016).

Obviously, the technological solutions presented in previous sections regarding the solid-state lighting, the networks of lighting fixtures, as well as the advances in sensors can certainly support the progress of the control approaches and methods. In other words, it is possible for control to significantly be shifted from simulation to emulation, allowing in this way the generation of new knowledge towards smart precision lighting in closed controlled horticultural environments. Therefore, various limitations, such as these reported by Van Ieperen and Trouwborst (2008), associated with the simulation of plants effects from inter-canopy lighting will be hopefully diminished in the near future.

6.5.3 Integration of Lighting Systems into Automation Control Systems

Lighting systems cannot be considered a single system that solely commands the lighting fixtures to produce the anticipated quantity and quality of light for specific crop cultivations. On the contrary, artificial lighting systems must be an integral part of the total automation and control system (Piromalis and Arvanitis 2016a). The ultimate system should include all the necessary sensors for light and plants growth sensors, as well as sensors for the environmental conditioning, together with specific electromechanical actuators and mechanisms for the control of heating and ventilation. Moreover, the automation system should be loaded with the necessary algorithms and procedures either in form of a computer software application or in the form of embedded firmware for microcontrollers and microprocessors, in order to be able to dynamically react to various influences and disturbances. Figure 6.23 illustrates an indicative block diagram of the integrated automation system.

At the same time, the automation system should be used as a data gateway in order to allow for remote control via the Internet or other communication media such as 3G/4G from mobile phone networks. Thereby, valuable technologies such as cloud computing and web applications in general could be exploited in sake of meeting the goals of smart precision lighting in horticulture.

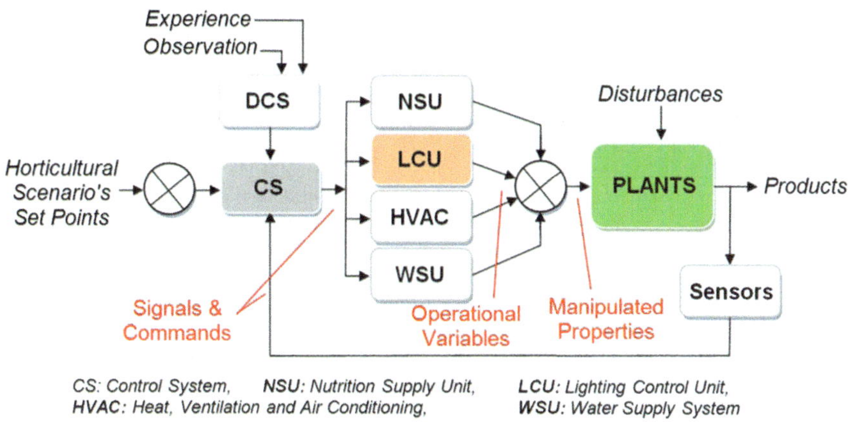

Fig. 6.23 The lighting system integrated into the overall automation and control system of the application

6.6 Conclusions

Although the benefits of stimulating and control of the various physiological properties and growth of plants through the use of artificial lighting have been proven during the last decades, artificial lighting is facing many challenges today, especially due to the fact that horticulture has become a rapidly changing sector. Closed controlled horticultural environments go beyond the boundaries of the established professional greenhouses and move to various urban deployments. On the other hand, there are significant technologies, either from other established application domains or new challenging ones, that can be beneficially integrated into the existing lighting systems for horticulture.

This work identified and presented specific requirements for lighting systems associated with horticulture in closed controlled environments, and it categorized them into functional, practical, electrical, and economical requirements. Moreover, a roadmap was proposed to achieve the smart precision lighting towards the implementation of artificial lighting applications of the twenty-first century. In particular, the suggested roadmap consists of three distinct steps, namely the exploitation of new lighting sources, the usage of the most effective networking technologies in order to have interconnected lighting fixtures, and the integration of intelligence into lighting systems. For each step, there is a thorough review and discussion, while design use-cases were also conducted. Ultimately, although horticulture is a traditionally conservative sector in terms of assimilating new technologies and it is prone to established solutions, it is now feasible, as presented in this work, to move into the future of smart precision lighting applications.

References

Albright LD, Both AJ, Chiu AJ (2000) Controlling greenhouse light to a consistent daily integral. Trans Am Soc Agri Eng 43(2):421–431

Association, LonMark Interoperability (2002) LONMARK® application layer interoperability guidelines. LONMARK Interoperability Association, 078-0120-01F, Version 3.3 October 2002. URL: http://www.lonmark.org/technical_resources/guidelines/docs/lyr733.pdf

Ayari O, Dorais M, Gosselin A (2000) Daily variations of photosynthetic efficiency of greenhouse tomato plants during winter and spring. J Am Soc Hortic Sci 125(2):235–241

Bergstrand KJ, Asp H, Larsson Jönsson EH, Schüssler HK (2015) Plant developmental consequences of lighting from above or below in the production of poinsettia. Eur J Hortic Sci 80(2):51–55

Böhme M, Grbic N, Paschko K, Pinker I (2015) Growth and internal quality of vietnamese coriander (polygonum odoratum lour.) affected by additional lighting with blue, red and green LEDs. In: De Pascale S, Jiang WJ, Connellan G (eds) Acta Horticulturae: International Society for Horticultural Science

Both AJ, Albright LD, Langhans RW (1998) Coordinated management of daily PAR integral and carbon dioxide for hydroponic lettuce production. In: Marcelis LFM (ed) Acta Horticulturae

Both AJ, Ciolkosz DE, Albright LD (2002) Evaluation of light uniformity underneath supplemental lighting systems. In: Dorais M (ed) Acta Horticulturae

Brazaityte A, Duchovskis P, Urbonavičiute A, Samuoliene G, Jankauskiene J, Kasiulevičiute-Bonakere A, Bliznikas Z, Novčkovas A, Bree K, Žukauskas A (2009) The effect of light-emitting diodes lighting on cucumber transplants and after-effect on yield. Zemdirbyste 96(3):102–118

Cho S, Dhingra V (2008) Street lighting control based on LonWorks power line communication. In: IEEE international symposium on power line communications and its applications, IEEE ISPLC 2008. Jeju Island, South Korea

Contreras S, Bennett MA, Metzger JD, Tay D (2008) Maternal light environment during seed development affects lettuce seed weight, germinability, and storability. HortScience 43(3):845–852

CREE (2016) Horticulture reference design, CREE Inc., June 2016, http://www.cree.com/~/media/Files/Cree/LED-Components-and-Modules/XLamp/XLamp-Reference-Designs/HorticultureReferenceDesign.pdf

CREE (2017) Cree XLamp XP-G3 LEDs, CREE Inc., CLD-DS139 Rev 0G, http://www.cree.com/~/media/Files/Cree/LED-Components-and-Modules/XLamp/Data-and-Binning/dsXPG3.pdf

DiLouie C (2006) Advanced lighting controls: energy savings, productivity, technology and applications. The Fairmont Press, Inc

Eigenbrod C, Gruda N (2015) Urban vegetable for food security in cities. A review. Agron Sustain Dev 35(2):483–498

Ferentinos KP, Albright LD (2005) Optimal design of plant lighting system by genetic algorithms. Eng Appl Artif Intell 18(4):473–484

Ferentinos KP, Albright LD, Ramani DV (2000) Optimal light integral and carbon dioxide concentration combinations for lettuce in ventilated greenhouses. J Agric Eng Res 77(3):309–315

Fierro A, Tremblay N, Gosselin A (1994) Supplemental carbon dioxide and light improved tomato and pepper seedling growth and yield. HortScience 29(3):152–154

Gómez C, Mitchell CA (2014) Supplemental lighting for greenhouse-grown tomatoes: intracanopy LED towers vs. overhead HPS lamps. In: Acta Horticulturae: International Society for Horticultural Science

Gömez C, Mitchell CA (2016) Physiological and productivity responses of high-wire tomato as affected by supplemental light source and distribution within the canopy. J Am Soc Hortic Sci 141(2):196–208

Han J, Huang GJ, Wu T, Sun XM, Liu Y, Gao RF (2015) Photoperiod affects morphology, flower productivity and photosynthesis of forced potted Paeonia lactiflora in greenhouses. In: Criley RA (ed) Acta Horticulturae: International Society for Horticultural Science

Harbick K, Albright LD, Mattson NS (2016) Electrical savings comparison of supplemental lighting control systems in greenhouse environments. In: 2016 ASABE Annual International Meeting: American Society of Agricultural and Biological Engineers

Hernández R, Kubota C (2016) Physiological responses of cucumber seedlings under different blue and red photon flux ratios using LEDs. Environ Exp Bot 121:66–74

Huynh TP, Tan YK, Tseng KJ (2011) Energy-aware wireless sensor network with ambient intelligence for smart LED lighting system control. In: 37th annual conference of the IEEE industrial electronics society, IECON 2011. Melbourne, VIC

IES, IESNA (2011) Lighting control protocols. In: Report IES-TM-23-11: Illuminating Engineering Society of North America—IES

Jeong SW, Hogewoning SW, van Ieperen W (2014) Responses of supplemental blue light on flowering and stem extension growth of cut chrysanthemum. Sci Hortic 165:69–74

Johansen NS, Vänninen I, Pinto DM, Nissinen AI, Shipp L (2011) In the light of new greenhouse technologies: 2. Direct effects of artificial lighting on arthropods and integrated pest management in greenhouse crops. Ann Appl Biol 159(1):1–27

Kitsinelis S, Kitsinelis S (2015) Light sources: basics of lighting technologies and applications. CRC Press

Kozai T, Niu G (2015) Challenges for the next-generation PFAL. In: Plant factory: an indoor vertical farming system for efficient quality food production. Elsevier Inc

Kyriakarakos G, Piromalis DD, Arvanitis KG, Dounis AI, Papadakis G (2015) On battery-less autonomous polygeneration microgrids: investigation of the combined hybrid capacitors/hydrogen alternative. Energy Convers Manag 91:405–415

LIN, Consortium (2006) LIN specification package, Revision 2.1

López C, Doval J, Pereira M, Pérez S, Dios J, López O (2007) DMX512 controller for high brightness RGB LED matrix. In: IEEE international symposium on industrial electronics, ISIE 2007. Caixanova, Vigo

Massa GD, Kim HH, Wheeler RM, Mitchell CA (2008) Plant productivity in response to LED lighting. HortScience 43(7):1951–1956

Maxfield C (2011) IPv4, IPv6, The Internet of Things, 6LoWPAN, and lots of other "Stuff"." EETIMES

Meng Q, Runkle ES (2015) Low-intensity blue light in night-interruption lighting does not influence flowering of herbaceous ornamentals. Sci Hortic 186:230–238

Morrow RC (2008) LED lighting in horticulture. HortScience 43(7):1947–1950

Newton S (2005) Art-Net and wireless routers. In: Asia-Pacific conference on communications. Perth

OSRAM (2015). Horticulture lighting with LEDs. In: OS SSL, NR AW: OSRAM Opto Semiconductors

Pinho P, Tetri E, Halonen L (2006) Synergies of controller-based LED drivers and quality solid-state lighting. In: PRIME 2006: 2nd conference on Ph.D. Research in MicroElectronics and Electronics. Otranto

Pinho P, Hytönen T, Rantanen M, Elomaa P, Halonen L (2013) Dynamic control of supplemental lighting intensity in a greenhouse environment. Lighting Res Technol 45(3):295–304

Piovene C, Orsini F, Bosi S, Sanoubar R, Bregola V, Dinelli G, Gianquinto G (2015) Optimal red: blue ratio in led lighting for nutraceutical indoor horticulture. Sci Hortic 193:202–208

Piromalis D, Arvanitis K (2016a) Precision lighting for controlled closed urban horticultural environments with emphasis on the use of LED technology. In: International conference on landscape and urban horticulture (LUH 2016). Athens

Piromalis D, Arvanitis K (2016b) SensoTube: A scalable hardware design architecture for wireless sensors and actuators networks nodes in the agricultural domain. Sensors 16(8):1227

Piromalis DD, Arvanitis KG, Sigrimis N (2013) DASH7 mode 2: a promising perspective for wireless agriculture. In: 4th IFAC conference on modelling and control in agriculture, horticulture and post harvest industry, AGRICONTROL 2013. Part 1 edn. Espoo

Piromalis D, Arvanitis K, Papageorgas P, Tseles D, Psomopoulos C (2016) LEDWIRE: a versatile networking platform for smart LED lighting applications using LIN-Bus and WSNs. Sens Transducers 200(5):50–59

Ponce P, Molina A, Cepeda P, Lugo E, MacCleery B (2014) Greenhouse design and control. CRC Press

Samuoliene G, Brazaityte A, Viršile A, Sirtautas R, Sakalauskaite J, Sakalauskiene S, Duchovskis P (2015) Photomorphogenetic effects in different plant life forms. In: Mauget JC, Godet S (eds) Acta Horticulturae: International Society for Horticultural Science

Simpson RS (2003) Lighting control: technology and applications. Taylor & Francis

Singh D, Basu C, Meinhardt-Wollweber M, Roth B (2015) LEDs for energy efficient greenhouse lighting. Renew Sustain Energy Rev 49:139–147

Smet KAG, Ryckaert WR, Pointer MR, Deconinck G, Hanselaer P (2012) Optimization of colour quality of LED lighting with reference to memory colours. Lighting Res Technol 44(1):7–15

Takagaki M, Hara H, Kozai T (2015) Micro- and mini-PFALs for improving the quality of life in urban areas. In: Plant factory: an indoor vertical farming system for efficient quality food production. Elsevier Inc

Taulavuori K, Hyöky V, Oksanen J, Taulavuori E, Julkunen-Tiitto R (2016) Species-specific differences in synthesis of flavonoids and phenolic acids under increasing periods of enhanced blue light. Environ Exp Bot 121:145–150

Tewolde FT, Lu N, Shiina K, Maruo T, Takagaki M, Kozai T, Yamori W (2016) Nighttime supplemental LED inter-lighting improves growth and yield of single-truss tomatoes by enhancing photosynthesis in both winter and summer. Front Plant Sci 7 (APR2016)

Thomas BA, Azevedo IL, Morgan G (2012) Edison revisited: should we use DC circuits for lighting in commercial buildings? Energ Policy 45:399–411

TI (2002) RS-422 and RS-485 standards overview and systems configurations, application report—SLLA070D. Texas Instruments

Ulrichs C, Mewis I (2015) Recent developments in urban horticulture—facts and fiction. In: Mauget JC, Godet S (eds) Acta Horticulturae: International Society for Horticultural Science

Van Ieperen W, Trouwborst G (2008) The application of LEDs as assimilation light source in greenhouse horticulture: a simulation study. In: International symposium on high technology for greenhouse system management, Greensys 2007. Naples

Van Iersel MW, Mattos E, Weaver G, Ferrarezi RS, Martin MT, Haidekker M (2016) Using chlorophyll fluorescence to control lighting in controlled environment agriculture. In: Lopez RG, Runkle ES, Currey CJ (eds) Acta horticulturae: International Society for Horticultural Science

Vänninen I, Pinto DM, Nissinen AI, Johansen NS, Shipp L (2010) In the light of new greenhouse technologies: 1. Plant-mediated effects of artificial lighting on arthropods and tritrophic interactions. Ann Appl Biol 157(3):393–414

Walerczyk S (2014) Lighting and controls. Fairmont Press

Wojciechowska R, Dugosz-Grochowska O, Koton A, Zupnik M (2015) Effects of LED supplemental lighting on yield and some quality parameters of lamb's lettuce grown in two winter cycles. Sci Hortic 187:80–86

Yurish SY, Cañete J (2013) High performance sensor nodes for wireless sensor networks applications. Sens Transducers 18(SPEC.ISS.1):92–99

Chapter 7
Edible Landscaping in Urban Horticulture

Mohammed Ibrahim Fetouh

Abstract Edible landscaping is simply a way of using veggies, herbs, fruits, and flowers that will perform multiple functions, such as for food, flavor, and ornamental appearance. Edible landscaping offers an alternative to conventional landscapes producing vegetables, fruits, and herbs for home use. Edibles can be independently grown or blend with ornamentals into existing yards and gardens into aesthetically pleasing designs. Many edibles function in landscape in just a perfect way, i.e., date palm is extensively used in landscape as a major tree, rosemary that is used as spice is a suitable for low-growing perennial hedge, and purple basil is perfect alternative for dark-leaved annuals for pots. The need for edible landscaping is gaining more interest due to the increasing of population and spreading of urban communities worldwide as it functions many goals leading us to explain the benefits and methods of applying edible ornamentals or as preferred to be titled as edible landscaping.

Keywords Edible landscaping · Edible ornamentals · Edible landscape designs
Vegetable gardens

7.1 Introduction

What is ornamental? Typically, plants that are grown for decorative purposes, but are not vegetables and fruits! Is there anyone who can deny the beauty of blooming stone fruits or ignore the decorative leaves of kale? This means that we can change the way we use them to double the benefit. Competition between edible plants and strictly ornamental is not always won by ornamentals. Many edibles have shapes, textures, flowers, and colors that can be attractive as ornamentals and play double role in landscape as well as a flavor to dinner table.

M. I. Fetouh (✉)
Department of Horticulture, Faculty of Agriculture, Tanta University, Tanta 31527, Egypt
e-mail: mohammed.fetouh@agr.tanta.edu.eg

D. Nandwani (ed.), *Urban Horticulture*, Sustainable Development
and Biodiversity 18, https://doi.org/10.1007/978-3-319-67017-1_7

Edible landscaping has gained popularity in recent years as gardeners and owners began to evaluate the return value for money spent on landscape. Since both ornamental and edibles share the same cost, maintenance, fertilizers and, above all, irrigation. Edible ornamental can be the perfect candidates output of the cost–value equation. With that in mind, expanding in edible landscaping requires looking for untraditional uses of edibles in landscape. Countless edible plants can partially substitute traditional landscape plants and easily integrate into a landscape and function in multiple ways.

Edible landscaping is not a new practice, but it is still a new trend in different countries. In fact, many gardens throughout history included both edibles and ornamental plants. Evidence of using edible ornamentals was recorded in ancient Egyptian temples. In here, we are discussing the benefits, and indicate the multiple functions performance of different edibles, such as for food, flavor, and ornamental appearance as alternative for common ornamental plants.

7.2 Why We Need Edible Ornamental in Urban Landscaping?

Edible landscaping has taking so many forms through centuries and was developed to serve one major purpose which is to secure food and medicine for urban citizen. Other benefits have shown up due to the extension in urban communities, There are several reasons that is why we shall shift to edible landscaping which can be sorted into four major categories—economic, environmental, and social.

7.2.1 Economic

Extension in urban communities was predicted to reach 60% of the world's population by 2030 according to United Nation's projections (2002). Urban landscaping, public gardens, or private gardens all are meant to be established to serve for people's comfort and luxury and keeping safe environment, but the foundation of such garden especially in public garden in urban communities or small cities require expensive infrastructure without a real direct benefits for public community. These gardens spaces are cut of farmland around cities that grows the food urban dwellers consume (Halweil 2002). That means the argent compensation of these cut off resources for the good of the population, providing food resources are becoming more critical every day even with the major development in food transportation and shipping.

The direct way for food supply is to grow personal food needs, communities, cities, or even individuals can contribute to their own food needs (Beck and Quigley 2003). The same resources used for public, urban, or private gardens maintenance and management could be easily used for food production using edible landscaping and meanwhile serves the purpose of landscaping which guarantees double benefits. Almost more than 60% of the common city block can secure around 40% of the community food supply (Halweil 2002).

Resources used efficiently in single edible landscapes found to show the same efficiency of a typical ornamental landscape. However, implementing edible landscape on large-scale residential areas shows more efficiency and become more productive (Beck and Quigley 2003). This means more food production in the community shrinking the charge for the shipping distance from farm to table. That represents the direct benefit of edible landscape linked to the chain of food production and consumption (Lovell 2010).

7.2.2 Environmental

Shifting to edible landscaping enables producing fresh foods as well as conserving many local environmental elements including: water, soil, wild life, pest control, energy conserving, and increasing the natural diversity around resident's homes.

These environmental and ecological benefits are considered as the same important as the production functions. Edible ornamentals in urban landscapes enable reducing energy consumption related to food production, transportation, packaging & processing, and waste management (Bohn and Viljoen 2005). Reducing energy consumption directly affects residential areas micro-climate through humidity regulation and shading which has its effect on global warming impact (Deelstra and Girardet 2000).

Edible landscapes are not just about providing food, but naturally if there is a food growing there will be wild life. This will include most of the natural food chain beginning with worms and insect which are in most cases the preferable meal for birds and a lot of animals. Edible woody landscapes consider a better choice for different kind of birds for nesting and reproduction which feeds on insects which help the reduction of pesticide in landscape, naturally (Henderson 1987). Mostly the habits of wild life changing seasonally mean different wild life will show up in the backyard or the frontyard on a cycle during the year giving the chance for bird watcher and wild life photographers to practice their hobbies not far from their windows.

7.2.3 Social

Edible ornamental is not just planting for food production, the benefits is beyond power consumption and multifunctional use for resources. Integration of urban agriculture into densely populated areas reconnects communities with a greater understanding of food; fosters diverse multi-cultural and inter-generational activities; food activities are linked to supporting communities (Dubbeling et al. 2009) in numerous ways shortly explained as follows:

- Integration of edible ornamental in the neighborhood instead of typical landscape considers favorable by resident in relation to freshness and food safety (Koont 2009).
- The food production process usually needs a lot of handwork even if it was in a form of edible ornamentals which provides more jobs especially in the low-income neighborhoods (Midmore and Jansen 2003).
- Increasing the chance for cooperation and socialization between neighbors and families (Holmer and Drescher 2009).
- Involving residents in the agriculture system increases their awareness regarding the environmental (East and Dawes 2009).
- Edible ornamental provide an easy applicable solution for common urban communities poor nutrition diseases as it provides direct access for high-quality fresh foods (Vitiello, 2008).
- The quality of nutrients for neighborhoods participant in edible landscaping will be better and the residents will have healthier diets as they will consume more vegetable and fruit in their diets (Parmer et al. 2009).
- Commercial production of foods is mainly for the major crops consumed by close consumers making it hard to access for rare foods. Edible landscaping gives the chance to experience rare foods growing by different residents (Wakefield et al. 2007).

To get all the benefits of from edible landscaping in urban communities, many efforts had to be considered while planning urban communities. The hardest challenge is to create an edible landscaping imitating an old natural habitat with multifunctional purposes meeting the needs for residents for favorable foods and does not harm the natural environment. In Table 7.1, Lovell (2010) summarized the various function of edible landscaping and their supportive strategies.

7.3 Edible Landscape Designs

Edible landscaping has no difference of common landscape, it follows the same roles and principles within the designing stage. The designer of edible landscape can count on it as sustainable and multifunctional project and take in consideration the community development programs, designing schemes, and housing type and design (Bhatt 2009).

Certainly, edible ornamentals can be integrated in urban agriculture in different designs at different scales providing vast benefits for the sake of urban communities. Edible landscape designer had to put in mind different point during planning and designing.

Table 7.1 Edible landscaping functions and their supportive strategies

Function	Description and justification	Supportive planning strategies
Production	Noncommercial production of local communities food categories Edible landscaping	Planning landscapes for sustainable production and resources use efficiency
Energy management	Local food production conserve power consumed during packing, cooling and transportation	Develop transportation systems for efficient food delivering system
Waste management	Recycling organic wastes for compost production and safety food fertilizing	Develop waste collection and composting system for local communities
Biodiversity	Landscape planning can support a various collection of native and imported plant species	Reshaping the distribution of landscape items to add more gardens and farms
Microclimate control	Landscape usually positively modify microclimate through humidity control, wind protection, and shade	Designing edible landscape to allow air movement to avoid climate worming conditions
Economic revitalization	Edible ornamentals offer more jobs in the low-income neighborhoods	Develop social network to introduce residents to new jobs
Community socialization	Gardening activity and sharing food products enhance socialization among residents	Integration of farming activity along with other social activity
Public health	Free access for fresh fruits and vegetables plus other green spaces usually enhance resident's health and physical activity	Explore opportunities to develop healthy lifestyle through community programming
Cultural heritage	Edible ornamentals can provide access to typical ethnic food that can be very rare for immigrant residents	Providing communities with immigrant population with community edible landscaping
Education	Edible landscaping enables kids and adult to know about food production, crops, nutrition, environment and other cultures	Organizing programs during summer and vacations for gardening activity

Design of multifunctional edible landscapes will depend on habits and nutrition requirement of residents, the design may include one or more form of edible plant designs. Edible landscape could be one of the next designs.

7.3.1 Vegetable Gardens

Vegetable gardening is an incident forms of edible ornamental even if it was not recognized under this term. Vegetable garden was the main supply for small families for fresh vegetable all the year. If it was organized in the appropriate way it serves as ornamental item.

Additional motivations for starting a vegetable garden is to secure rare and ethnic foods that cannot be easily found in stores. Vegetables may have undesirable decorative structure, but still could form a desired decorative and productive form when arranged in a suitable way in soil beds or grown in container as well.

For successful vegetable garden establishment, there are many considerations that had to be taken in mind to maintain vegetable crops during season, i.e., site-specific growing conditions. Vegetables mostly require fertile well-drained soils and adequate direct sun exposure for at least 6 h. Climate is another limiting factor for vegetable selecting and cultivation that includes growing season duration, first and last frost dates as well as the growing season average temperatures.

7.3.2 *Edibles Integration in Ornamental*

Colored or variegated vegetable plants and flowering fruit trees can be used alone or along with each other's or even with ornamental plants to design a distinguish landscape. The landscaper applies all the main the roles of landscape when replacing ornamentals with edibles. Many edible plants serve the same function as well as ornamentals. Untraditional use of edibles as ornamentals enhances landscapes by adding additional use of landscape items beyond just good-looking plants. Some edibles are good as ornamentals in making hedges or in basket or containers cultivation as good as ornamentals.

7.3.3 *Edible Woody Landscapes*

Edible landscape using woody edible shrubs and trees is the most common and easier to manage form of edible landscape. Most of well-known fruit shrubs and trees exhibit extraordinary flowering as well as decorative leaf. Fruit trees such as cherries, pears, apples, raspberries, peaches, and even date palm are common sights in gardens in different country. Some of this fruit crops are big enough to be used as a shade tree or can be dwarfed to grow in a container. Edible shrubs are appropriate for creating fences or small hedges. Woody edible landscaping are more suitable for garden in arid and semi-arid region.

To design a well-functioning edible woody landscape, designer have to consider:

- **Planting Objectives**: Is there any planning for product harvest and processing or it will be left for wildlife?
- **Wildlife**: The edible ornamental type will define the kind of animals or birds that will be attracted to it (e.g., songbirds, squirrels)? That means that the designer can define the types of birds and animals that will be a part of the design later.
- **Space**: The available space determines the type and the design of edible space. The available space measurement should take in consideration any available space along with horizontal yards that mean calculation of shrubs, trees and climbing vines, so multi-levels of edible gardening could be created.
- **Time**: The expected time to receive the edible products will define which type of edibles to grow. Herbaceous and vegetable plants usually produce their products sooner than shrubs and trees. Other woody ornamental can take several years to products and may be take more in case of some palm types.
- **Viewability**: One of the main concepts of landscape is to get useful of nearby gardens as well as wild life. So, when designing an edible landscape it should be considered.
- **Diversity**: Planning edible landscape should make certain of plant diversity in species, shape, size, and growth habits to insure continues flowering and production during the year and maintaining the landscape function all the time as possible.
- **Maintenance**: selecting landscape item defines the required maintenance. Many improved selections require little maintenance once established.
- **Beauty**: The main purpose of landscape is to create and imitate natural beauty of plants. So, the selection edible plants to create a landscape should aim to obtain the same function and in the same time enable attracting wild life into the garden.
- **Plant Suitability**: Selected trees or shrubs for landscapes should be suitable to the growing site at their mature growth stage, resistant to the site weather, insect and diseases. In this concern, native species are more preferable.
- **Availability and Price**: Most of edible shrubs and trees are available in nurseries and stores with reasonable prices especially native plants. In some case, certain plants are hard to find or require special delivery. The project budget and species availability will define what plants to be selected for planting along with the desires of the residents.

7.4 Alternative of Typical Landscape Plant Categories

Edible landscape designer had to list a number of edibles as alternatives to typical landscape plants before designing edible landscape. Such knowledge of alternative use of plant is the principle of edible landscaping. In here we are going to list a number of edible plants that can function as a different ornamental plant category.

Alternative Functional of edible plants as ornamentals.

7.4.1 Ground Covers

Strawberry: Fragaria species

Description: a small herbaceous plant with decorative leaves and white flowers usually flower in the spring.

Habitat and Distribution: Strawberries prefer to grow in the North Temperate Zone with distribution in different region of the world, sometimes it grows in the top of the mountains in the southern part of the world.

Edible Parts: The preferable fleshy and sweaty fruits are consumed fresh or processed as a source of vitamin C.

Landscape function: Strawberry can be used in landscape in different ways including planting as ground cover for slopes, as basket plants and vertical planting.

Menthe species

Description: Short and small creeping herbaceous plants with opposite hairy leaves known with its sweet fragrant smell. Flowers in most case are white or pink.

Habitat and Distribution: This small plant is native of Europe and distributed world wide with various number of verities and it can be found in any continent of the world.

Edible Parts: Leaves are consumed fresh or dried and boiled to make a tea for its volatile oil.

Landscape function: Mint consider a perfect ground cover as it grows very quickly and recover its growth after several cuts during the year especially at the warm months. It is also suitable for containers and baskets. It also grows good in wetlands.

7.4.2 Annual and Perennial

Asparagus: Asparagus officinalis

Description: Perennial plant looks like the ornamental asparagus on growth at mature stage. In the spring it usually begins the growth cycle of new growth of buds in the shape of cluster of a green finger.

Habitat and Distribution: Asparagus prefers the temperate zone of the world but still known and preferred in different part of the world.

Edible Parts: The young fleshy stems (fingers) are edible as long as they still immature, once it grows up it will not be good for cooking. Asparagus fingers are only edible cooked.

Landscape function: Asparagus can be grown in garden beds and garden borders as it resembles the growth of ornamental asparagus.

Artichokes: *Cynara scolymus*

Description: Artichokes are perennial, unlike many vegetables, and will return each spring. It is one of the biggest vegetable in plant size. The green leaves reach around 80 cm in length. The flower bud grows up to 15 cm diameter and usually picked before opening.

Habitat and Distribution: Artichoke is cultivated worldwide including the Mediterranean, France, Italy, and Spain. It grows best in full sun exposure, moist, and sandy environments.

Edible Parts: The heart of the flower buds is the only edible part of artichoke. The heart is the internal fleshy base of the flower bud. The florets in the middle are inedible.

Landscape function: Artichoke plants are increasingly being used for their ornamental value due to their large leaves and vibrant purple flowers. The best use of artichoke is to blend with annual flowers or small perennial in flower beds and containers adding color contrast and texture.

Sage: ***Salvia officinalis***

Description: Sage is a well-known perennial evergreen small shrub. The stems are semi-woody with opposite grayish hairy leaves. Plants produce spikes with blue to purple flowers in the winter.

Habitat and Distribution: Sage belong to family Lamiaceae, originated in the Mediterranean and from there it was distributed to the world.

Edible Parts: Sage leaves have a strongly aromatic volatile oil and are frequently used fresh or dried in cooking as a flavor or boiled as a tea.

Landscape function: Sage plants are widely used in gardens for their ornamental value due to their leaves and flowers. Sage can be mixed in flower beds or can be grown as a small biennial hedge.

Rosemary: ***Rosmarinus officinalis***

Description: Rosemary is an evergreen perennial shrub, generally erect, rounded, with aromatic gray-green leaves and purple flowers.

Habitat and Distribution: As a member of the Lamiaceae family, Rosemary originated in the Mediterranean region, from where it has been introduced into all continents. The plant considered a heat and drought-tolerant and easily adapts to boor soil conditions.

Edible Parts: The fragrant small leaves are consumed fresh or dried for different purposes. It is also used as an herb to savor meat, savory dishes, and salads.

Landscape function: Since it is drought-tolerant, the plant, the plant has a long history in landscape as small hedge, flower border and it can be single grown in outdoor containers.

Chamomile: ***Matricaria chamomilla***

Description: Chamomile is distinguished as annual plant with erected smooth and fragile stems with pinnate leaves. The plant plums in the late winter and early summer depending on the weather condition and growing cite. The flower has a strong aromatic smell white in color for the outer part and yellow for the disk florets.

Habitat and Distribution: Chamomile grows naturally near population areas in Europe, Asia, and North America. It is considered as a weed as the seeds widely spread with the winds and easily germinate in early winter.

Edible Parts: Flowers is the main edible part consumed fresh or dried for making chamomile tea. Flowers are used in herbal remedies as an anti-inflammatory and antiseptic.

Landscape function: Chamomile can be grown as a flowering annual in flower beds or mixed with other shrubs.

Sweet basil: *Ocimum basilicum*

Description: Sweet basil is an aromatic perennial plant. Stems and leaves are almost hairless with opposite ovate green leaves and purple in Dark Opal variety. The flowers are white or pink.

Habitat and Distribution: This species has been widely cultivated for thousands of years, and hence its exact origin is obscure. It is thought to have its origins in India and from there it was distributed to Europe and the Middle East and later it was introduced to America.

Edible Parts: Aromatic leaves usually consumed fresh and not suitable for cooking recopies. Fresh leaves is a main constituent in green salad in many places as a flavoring.

Landscape function: Sweet basil green or purple is a suitable plant for pruning to small hedges, border, or grown as container specimen.

Kale: *Brassica oleracea* var. *acephala*

Description: Kale is a Biennials herbaceous plant treated as an annual plant with different varieties of different colored leaves. It is still firmly edible but growers value it more as an annual ornamental.

Habitat and Distribution: It is difficult to define the origin of this species, but it is well-known as it was first introduced for cooking by Greeks and Roman, but it is considered native of Mediterranean.

Edible Parts: All kale varieties are edibles, even the strictly ornamental ones, ornamental kale have more bitter taste if consumed raw and the bitter taste is removed when cooked.

Landscape function: Kale is a perfect bedding plant usually used for creation of colored beds at different shape or it can be used as a single plant in containers. Ornamental kale has a wide range of colors and considered a perfect choice in the winter.

7.4.3 Shrubs

Blackberry: *Rubus allegheniensis*

Description: Blackberry is a famous wood shrub grown as vines in a biennial pattern as it produce leaves in the first year of growth and produces fruits on the second one; then it dies followed by regrowth of new canes in the next season. It belongs to Rosaceae family with the typical trifoliate leaves and white flowers during early summer.

Habitat and Distribution: This plant grows in wild areas as well as a cultivated crop in the temperate regions. It grows better in light shade and rich soils.

Edible Parts: The only edible parts are the fleshy sweet fruits famously good in pie and other deserts. The frits turned to black at full ripe stage.

Landscape function: The plants are attractive during flowering and fruiting time. Blackberry canes can be used for building fences on wires for dividing areas.

Red Raspberry: *Rubus idaeus*

Description: very similar to blackberry except of leave structure, new canes usually carry compound leaves of five to seven leaflets compared to three leaflets for blackberry. The canes do not die after the second year of growth and instead it produces a large number of lateral branches.

Habitat and Distribution: Red raspberry is very similar in growth habits to blackberry and it grows in a wide range of temperate regions from North America to Canada.

Edible Parts: The fruits look identical to blackberry except it is red at the final stage. The fruits are a group of aggregated droplets around the core.

Landscape function: According to its growth habits in the wild it can be used as a single shrub or in a group in shady areas under an old tree with extended canopy.

Blueberry: *Vaccinium corymbosum*

Description: One of the famous edible shrubs in North America. Considered as semi-evergreen small, erected to large shrubs. It blooms a white to bale binky bell-shaped flowers.

Habitat and Distribution: Vaccinium genus is mostly distributed in Europe and North America. These plants prefer open, sunny areas.

Edible Parts: The edible part is the berry blue fruits which is consumed fresh or processed.

Landscape function: blueberry's erected shrubs are ideal for low hedges establishment.

Pomegranate: *Punica granatum*

Description: Pomegranate grows up to be considered a deciduous big shrub or a small tree. The glossy narrow leaves arranged in an opposite position on the new branched formed in the early spring and later it forms big red flower with three to seven petals that blooms in a wide range of time during the summer.

Habitat and Distribution: The pomegranate originated in the regions extended from Iran to India. In the meantime, it is cultivated in different region of Mediterranean, Africa, and southern Europe. The plant is well known for drought and salinity tolerance.

Edible Parts: Pomegranate has a unique rounded berry fruit with thick, radish skin containing over 1000 seed. The edible part is the sarcotesta tissues coating the seeds with red or purple color.

Landscape function: The edible pomegranate is a perfect choice for cultivation in bad soil condition and adapted to drought and salinity and in the same time it produces a large red flower for extended season. The flower of edible pomegranate is the same as in the ornamental varieties that does not develop to produce fruits.

7.4.4 Trees

Cashew: ***Anacardium occidentale***

Description: Cashew is a large tree with extended evergreen canopy that functions perfectly as a shade tree. The tree could reach 12 m with yellow flowers develop to a fruit with a particular structure.

Habitat and Distribution: Cashew nut is considered one of the old-world plants that was originated in India and spread cultivation to different parts of the world and still grows wild in some places in India and Africa.

Edible Parts: Cashew is one of the well-known nuts and the fruit itself is quite edible and contains one seed that is famous to be roasted before eating.

Landscape function: As a spreading, evergreen tree, cashew is grown single as a major shading tree, fruits shape, and color also is attractive.

Sugar apple: *Annona squamosa*

Description: An evergreen small tree grows up to 6 m, multi-branched with simple dark green leaves. The tree produces a rare green color fruit.

Habitat and Distribution: It is thought that sugar apple was originated in India, but it can be found in most of tropical and subtropical region of the world. Although it grows in tropical regions, it is considered a cold tolerant plant.

Edible Parts: Fresh fruit's flesh are eaten raw or used in different kinds of processing for juice rich in calcium, phosphorus, and iron.

Landscape function: considering the small size of the tree, it can be grown in a line for separating areas. Or it can be single grown over lawns.

Mango: *Mangifera indica*

Description: Mango is a well-known evergreen tree that usually blooms in the spring with a massive number of flower cluster that develops to a remarkable fruit shape that differs in shape and colors according to the varieties.

Habitat and Distribution: The tree is native of India and grow in the worm parts in the world and grow in different kind of soil, but very susceptible to cold and frost.

Edible Parts: Mango fruits is one the most famous fruits known all over the world for its juicy sweet flesh that is consumed fresh or processed for juice.

Landscape function: Mango is one of the most favorite fruit and also the bright leaves make it a suitable choice for single tree or it can be grown in groups. When it reaches full maturity the tree is considered a perfect shading tree.

Mulberry: *Morus alba*, *Morus nigra*

Description: A deciduous trees retrieve growth in early spring producing large-lobed green leaves. The flower develops quickly to white or black fruits.

Habitat and Distribution: Mores trees are adapted to a different temperate region of Europe, America, Asia, and Africa and can be found wild in any frost.

Edible Parts: Mulberry fruits are very sweet and usually eaten fresh.

Landscape function: Mulberry is a deciduous tree, so it is suitable for growth at the garden edge away of swimming pools. It could be used as shading tree in summer and to enjoy sun in winter.

Papaya: ***Carica papaya***

Description: A small evergreen palm like tree with leaves around the apex. Flowers develop to green fruits grows directly on the trunk and turn yellow when ripe.

Habitat and Distribution: Papaya grows in different climate condition from temperate region to tropical and subtropical. It prefers wet lands and sunny places.

Edible Parts: The fruit edible part is the yellow flesh that is consumed fresh or cooked for its rich content of vitamin C.

Landscape function: Papaya has a palm-like structure that makes it a good candidate in case of planting single trees in the center of flower bed, or to be grown in group in a corner and sometimes it could be grown in a straight line.

Hazelnut: Corylus species

Description: A small deciduous bush grows up to 3 meters except for one species on Turkey and other one in China that are considered a large tree.

Habitat and Distribution: This genus of hazelnut is considered natives of North America and can be found in the eastern cost of America to Canada. It is also very familiar in Europe and Asia. It is adapted to a wide range of soils and prefers to be well drained.

Edible Parts: The small nut is edible and can be cracked to eat in the autumn. The nuts have a high nutrition value.

Landscape function: This small bush is considered fast grown, so it is suitable to be grown as fences or to be grown alongside with tree lines.

Olive: *Olea europea*

Description: Olive trees are small evergreen tree, squat and tend to be vertical growth in reach. The leaves are oblong small silvery green.

Habitat and Distribution: Olive tree is native of the Mediterranean, prefers sunny areas and good soil. The tree considered highly tolerant for salinity and drought.

Edible Parts: A tea can be made from the leaves. The fruits are used for oil extract and pickling.

Landscape function: Olive trees are considered a salt and drought-tolerant which makes it a perfect tree in arid regions. The tree can be pruned into formal shapes.

Citrus: species

Description: Citrus species grow to different sizes depending on the specie. All species are evergreen, typically aromatic flowers solitary or in small corymbs. The fruits are usually orange except for lemons which are green or yellow.

Habitat and Distribution: Citrus species is now growing all over the world. It is originated in Asia and it can be found in all the Mediterranean and from there it was spread all the world. It prefers a temperature range from 23 to 30 °C to grow and root zone temperature over 12 °C.

Edible Parts: Citrus is commercially grown for its fruits rich in nutrition value especially vitamin C. Some cultivars are used for volatile oil distillations. Fruits can be processed for different king of product.

Landscape function: Due to the flower's strong aroma and refreshing fruits, citrus trees grown in containers was a part of historical gardens through the seventeenth and eighteenth centuries. Until now, most of the private gardens have to contain at least one of citrus species and in most case owners prefers lemon trees. Trees can be grown in containers, single of in groups.

Peach: ***Prunus persica***

Description: Small deciduous tree grows up to 4 m tall with lanceolate, pinnately veined leaves. The tree blooms in early spring, flowers blooms before leaves. The flowers have a unique pink color appearing in solitary or in pairs.

Habitat and Distribution: Peaches are native of China then it was transplanted in Persia, Europe, and Africa. Peaches prefer cold winters regions and most cultivars require around 500 chilling hours.

Edible Parts: The fruits are consumed fresh and for making juice and jams. The fruit is usually whitish or yellow with a sweet taste.

Landscape function: Peach trees are favorite in landscape for their pink flowers that appears on wood before leaves. It is also suitable for containers and it is preferred by Bonsai Hobbyists.

Loquat: ***Eriobotrya japonica***

Description: Loquat is a subtropical evergreen fruit tree has a short trunk and a rounded to upright canopy. Leaves are large fuzzy, and the tree blooms a fragrant cluster of flowers that develop to yellow fruits ripe in early spring.

Habitat and Distribution: Loquat is native of southeastern and central China and commercially grown in Mediterranean as well as different part of the world.

Edible Parts: Fruits vary in sweetness and flavors, but all varieties usually consumed fresh and can be cooked to different recipes of deserts and jams.

Landscape function: Loquat trees are a good choice for lawn trees, with a glossy leaves and a naturally attractive shape. The small tree is also adequate for container planting.

7.4.5 Palms

Edible palms function typically as ornamental palm in landscape, thus using productive palms in landscape has been used for centuries. Using edible palm will differ from region to another depending on the availability of palm. Besides, palm has another benefits rather than using in landscape or for productive use.

Sugar palm: ***Arenga pinnata***

Description: Very common Asian feather palm tree grows up to 15 m and the leaves reach up 6 m long. Flowers form in a huge cluster below the leaves to develop the fruits later.

Habitat and Distribution: This kind of palm is native of Asia and east India and is very popular in Malaysia, Indonesia, and the Philippines as it prefers tropical regions.

Edible Parts: Sugar palms are a multipurpose species and the fruit is mainly harvested for its sugar contents and the heart of the apical tip is edible as many other palm kinds. Also, the sap of the tree can be collected as a juice.

Sago palm: ***Metroxylon sagu***

Description: One of the most famous feather palm in the tropical rain forests. It is a small palm, and grows up to maximum of 9 m. The main core of this palm is a starch with the minimum contents of fibers.

Habitat and Distribution: Sago palm is native of tropical Sothern Asia. It prefers wet lands, lakes, river banks, and swamps shores to grow.

Edible Parts: The core of the trunk is the edible part for this kind of palm. The high content of starch makes it master food for the resident in its native habit. One trunk is thought to be enough to feed one person for a year.

Coconut: ***Cocos nucifera***

Description: A feather palm with narrow trunk, leaves usually aggregate at the top and the dead leaves fall by itself leaving a clean trunk. Flowers are formed in clusters between leaves producing the famous coconut fruits.

Habitat and Distribution: Coconut is the most famous palm tree growing on the shores of the tropics.

Edible Parts: The coconut fruits are consumed fresh for its rich milk which is rich in vitamins and considered a complete diet. The ripe fruits usually left to completely dry to extract coconuts.

Date palm: *Phoenix dactylifera*

Description: The most famous feather palm tree known all over the world. It is famous for its unique trunk, fruits, and the huge variety diversity. The trunk diameter and high as well as leaf length and structure vary among varieties.

Habitat and Distribution: Date palm is native of North Africa and Arabian part of Asia and from there it reaches different regions of the world and today it can be found in Indonesia and USA. It grows best in the arid semi-tropics.

Edible Parts: Date palm contains a huge diversity of fruit types which is a sweet taste and some varieties are consumed fresh at the ripe stage and some other varieties must be dried first to condense its sweetness. All kind of date palm fruits cannot be eaten before ripe.

7.4.6 *Cacti*

Prickly pear cactus: Opuntia species

Description: Opuntia is famous for its flat pad like transformed stems with sharp pointed thorns. Some of the new varieties does not have thorns.

Habitat and Distribution: A member of Cactaceae family found to be dominant in arid and semi-arid regions adapted to dry sandy soil conditions and it can be found in any desert.

Edible Parts: The main edible product of opuntia is the fruits which are the main product, the young pads cooked as a vegetables especially non-thorn varieties. The seeds contain a highly price oil when extracted.

Landscape function: Opuntia is used for rock garden and sandy cliffs as it is adaptation to arid conditions. Once it was established in the soil, it might not need watering any more.

7.4.7 Aquatic and Semi Aquatic Plants

Lotus: ***Nelumbo species***

Description: Lotus is an ancient strictly ornamental plant and was used in the ancient Egyptian temples. All nelumbo species are an aquatic plant growing in river banks and ponds with a huge floating leaf reaches around 1.5 m diameter and in other varieties leaves grow above water surface. Flowers are pink or white.

Habitat and Distribution: The first-known origin for the pink species was ancient Egypt several thousand years ago. The Egyptians consider it a holy plant and was cultivated in temples ponds and in the holy river Nile banks in the shallow waters. Yellow lotus is native of North America. All ancient cultures grew lotus in their water gardens, which include Egyptian, Chinese, and Japanese.

Edible Parts: The main edible part is the underwater rhizomes which is rich in starch. Also, the young leaves could be cooked as a vegetable.

Landscape function: Nelumbo is mainly used in landscape in ponds for their attractive leaves and flowers and it has been used by ancient Egyptian, Japanese, and Chinese.

Water lily: ***Nymphaea odorata***

Description: Water lily is an aquatic plant with large triangular floating leaves. The flowers are weight with fragrant smell. Flowers are red or white. The plants are stable in the mud due to fleshy thick rhizomes.

Habitat and Distribution: Water lilies are native of tropical and subtropical regions and can be found wild in most of the known rivers of tropics.

Edible Parts: Most of the plant part is edible, flowers, seeds are edible cooked or raw, rhizomes are rich in nutrition content can be cooked or dried and grind to a flour. Also, the seeds can be grind to flour as well.

Landscape function: ***Nymphaea*** is another example of ornamental plants that can be edible and like Nelumbo is used in landscape in ponds for their attractive leaves and flowers.

Canna lily: ***Canna indica***

Description: Canna lily is a semi-aquatic perennial plant that grows from underground rhizome to produce false stem formed from leaf stalk. Leaves are large green or variegated. Flowers are usually red, orange, or yellow.

Habitat and Distribution: Canna grows wild in wetland and beside river banks in tropical.

Edible Parts: Canna rootstock is full of starch and completely edible.

Landscape function: Canna is a typical landscape plant used for their glossy leaves and attractive flowers, plant used in flower beds or grown between tree lines.

7.4.8 Vines

Hops: *Humulus lupulus*

Description: Deciduous perennial herbaceous vine that grow very quickly with scented flowers. The growth begins in early spring and the new shoots support itself if it does not find something to climb.

Habitat and Distribution: It belongs to family Cannabaceae that originated in Europe and North America. The shoots die back in the winter and resume growth in the spring. It prefers rich soil and sunny places and it needs a strong support material to climb above it as it is vigorously grow.

Edible Parts: The golden hop vine has been bred for flower production and brewing beer.

Landscape function: Hops vines are used as a traditional vine covering arching and pergolas.

Kiwi: ***Actinidia deliciosa***

Description: one of the most famous edible vines that can be used in landscape, the plant is considered a small spreading or climbing shrub with deciduous long-petioled leaves. The flowers are a small white fragrance appearing in single or grouped in three in the leaf axil and later develops to a small fuzzy brown oblong fruit.

Habitat and Distribution: Kiwifruits are native vines of southern China and considered as the national fruit.

Edible Parts: The edible fruits are famous for its green juicy flesh with a quite acid taste known to be rich in vitamin C.

Landscape function: The deciduous vine will be a perfect alternative for typical landscape climber plants.

References

Beck TB, Quigley MF (2003) Emergy evaluation of food productioin in urban residential landscapes. Urban Ecosyst 5(3):187–207

Bhatt V, Farah LM (2009) Designing edible landscapes. Open House Int 34:5–7

Bohn K, Viljoen A (2005) More space with less space: an urban design strategy. In: Viljoen A (ed) CPULS: continuous productive urban landscapes—designing urban agriculture for sustainable cities. Elsevier, Amsterdam, The Netherlands, pp 11–16

Deelstra T, Girardet H (2000) Urban agriculture and sustainable cities. In: Bakker N, Dubbeling M, Gundel S, Sabel-Koschela U, de Zeeuw H (eds) Growing cities, growing food: urban agriculture on the policy agenday. Deutsche Stiftung fur Internationale Entwicklung (DSE): Feldafing, Germany, pp 43–65

Dubbeling M, Bracalenti L, Lagorio L (2009) Participatory design of public spaces for urban agriculture, Rosario, Argentina. Open House Int 34:36–49

East AJ, Dawes LA (2009) Home gardening as a panacea: a case study of South Tarawa. Asia Pac. View. 50:338–352

Halweil B (2002) Home grown: the case for local food in a global market. World watch Institute, Danvers, MA

Henderson CL (1987) Landscaping for wildlife. Minnesota Dept. of Natural Resources, St. Paul, MN

Holmer RJ, Drescher AW (2009). Allotment gardens of Cagayan de Oro: their contribution to food security and urban environmental management. In: Knie C (ed) Urban and peri-urban developments—structures, processes and solutions. Southeast Asian-German Summer School Program, Cologne, Germany, pp 149–155

Koont S (2009) The urban agriculture of Havana. Mon Rev 60:44–63

Lovell ST (2010) Multifunctional urban agriculture for sustainable land use planning in the united states sustainability 2:2499–2522. https://doi.org/10.3390/su2082499

Midmore DJ, Jansen HGP (2003) Supplying vegetables to Asian cities: is there a case for peri-urban production? Food Policy 28:13–27

Parmer SM, Salisbury-Glennon J, Shannon D, Struempler B (2009) School gardens: an experiential learning approach for a nutrition education program to increase fruit and vegetable knowledge, preference, and consumption among second-grade students. J Nutr Educ Behav 41:212–217

United Nations Future World Population Growth to be Concentrated in Urban Areas of World According to New Report Issued by United Nations Population Division POP/815 21 March 2002 Retrieved from http://www.un.org/esa/population/publications/wup2001/WUP2001pressrelease.pdf

Vitiello D (2008). Growing edible cities. In: Birch EL, Wachter SM (eds) Growing greener cities. University of Pennsylvania Press, Philadelphia, PA, USA

Wakefield S, Yeudall F, Taron C, Reynolds J, Skinner A (2007) Growing urban health: community gardening in South-East Toronto. Health Promot Int 22:92–101

Chapter 8
Urban Landscaping in the Arid Zone

Suresh Kumar and R. N. Kulloli

Abstract Hot arid zones with their climatic extremes, poor shallow soils, deep, often saline water, and unpredictable weather are stressed areas for vegetation to survive and grow, more so in respect of garden plants. With additional care, however, gardens in such areas can have all the components. Accordingly, efforts in the last 35 years on testing the adaptability and suitability of a large number of ornamental species have resulted in selecting different types of trees, shrubs, hedges, creepers, Bougainvilleas, palms, shade-loving plants, bulbous plants, cacti, and succulents as mentioned in this paper for urban landscaping in the Indian desert.

Keywords Arid lands · Urban landscaping
Bougainvilleas · Cacti · Shrubs · Hedges · Climbers

Hot arid regions are known for their hot climate, scarcity of water, and scanty vegetation. Hot gusty winds in most part of the year and frosty nights during winter further aggravate the harshness of weather. Yet, man has colonized these lands and in their own ingenious way, also did landscaping around their homes, villages, town, markets, and other such places. In this process, a whole lot of plant species have been found suitable for people as also for the researchers. Jindal (1970) gave an account of flowering shrubs of India covering only trees and shrubs. "Ornamental horticulture in India" described in detail about eight major ornamental crops emphasizing their selection, hybridization, and agro-techniques (Chadha and Choudhury 1986). Trevedi (1987) in a book on home gardening gave basics of gardening.

S. Kumar (✉) · R. N. Kulloli
Central Arid Zone Research Institute, Jodhpur 342003, Rajasthan, India
e-mail: sk_ecology@yahoo.co.in

S. Kumar
Central Arid Zone Research Institute, 124, Subhash Nagar, Pal Road, Jodhpur 342 008, India

R. N. Kulloli
Botanical Survey of India, Arid Zone Regional Circle, Jodhpur 342 008, India

D. Nandwani (ed.), *Urban Horticulture*, Sustainable Development and Biodiversity 18, https://doi.org/10.1007/978-3-319-67017-1_8

Sabina (2009) described ornamental trees, shrubs, foliage plants, climbers, bulbous plants, succulents and cacti bromeliads ferns and fern allies, ornamental grasses, sedges, rushes, and water plants but there was little for hot arid agroclimate. The latest work "Hand Book of Horticulture" by Chadha (2015) is again a monumental work covering almost all horticultural crops from production point of view. Thus, information on urban landscaping in hot arid areas is very scanty and scattered. Though basics of landscaping remain, by and large, same in arid areas as elsewhere, a few adaptive measures can make urban landscaping a successful enterprise in hot arid regions. These details along with suitable species of trees, shrubs, and other foliage plants found suitable in Indian arid zone based on my experience of 35 years of managing such gardens are detailed in this chapter.

8.1 The Basics

In order to ensure success of landscaping in arid areas, following issue needs prior scrutiny:

1. Availability of at least one meter deep good soil.
2. Perennial source of water for irrigation.
3. Provision of protection.
4. Aspect of the open area proposed for landscaping whether it will be receiving sunlight whole day if south facing or part of the day, if east or west facing or, no direct sun if north facing.
5. Proper provisioning of drainage in both normal and abnormally high rainfall.

The emphasis on one or all of these will be further determined by the location and size of landscape area. The variety of locations in an urban landscape include the following:

1. Hospitals and public dispensaries.
2. Bus stand and railway station including other public parking areas.
3. Parks and gardens visited by larger public, playgrounds, and stadia.
4. Marriage places.
5. Religious places: temples, mosque, and gurdwara.
6. Linear areas like highways, roads, railway line siding, avenues, and streets.
7. Water bodies like ponds, lakes, stream banks, and canal banks.
8. Schools, colleges, university campus, and places of education.
9. Historical and ancient places like forts, palaces, or other memorials.
10. Markets, bazaars, shopping complexes, and malls.
11. Institutional complexes.
12. Home gardens.

The selection of species at any of the above sites will be based on following criteria:

1. Height of tree, shrub, and undershrub.
2. Shape and size of the canopy to assess it shading.
3. Color, bunchiness, longevity, density, and luster of foliage (Evergreen or Deciduous).
4. Color and fragrance of flowers.
5. Phenology of the species.
6. Planting patterns and geometry requirement.
7. Growth rate and lopping/shaping periodicity.
8. Space available from permanent structure like wall side, road paving side, etc.
9. Issues related to allergenic potential of spp.
10. Social, ethnic, and religious needs.

8.2 Preparation for Plantation

Pre-plantation preparation of land includes bush cleaning and leveling. Thereafter, it should be deeply plowed and worked up to expose soil for solarization so as to kill all fungal/bacterial or nematode infestation. An irrigation prior to plantation will result in regeneration of seed bank and these regenerated saplings are actually weeds which should be removed forthwith. Thereafter, the required layout should be given.

As a rule, the soil modification of the plantation area or pit will have to be carried out. If the soil is predominantly sandy, the potting medium will have a mix of local soil, farmyard manure, and pond silt in the ratio of 1:1:1. In case of clayey soil, the medium comprises local soil, farmyard manure, and river sand in the ratio of 1:1:1. The use of water-retaining gels, provision of drips and sprinklers, and covering with shade nets are decided on the basis of local needs and resources availability. In case irrigation water is saline, the application of gypsum may also be needed, the quantity depending upon the level of salinity. Regular watering, weeding, shaping, cutting, and removal of dead trees and shrubs as well as dried seasonals are regular activities.

8.3 Species for Landscaping

Species suitable for landscaping in hot arid urban areas include trees, shrubs, hedges, creepers/climber, bougainvilleas, shade-loving plants, bulbous plants, palms, cacti, and succulents.

Fig. 8.1 *Agave weberi* and hedge of *Clerodendrum* sp.

8.4 Designing Landscapes

This is determined by the availability of space, distance from wall/permanent structure, and direction to which it faces sun. Normally, shrubs of varying size are preferred in masses or clusters to give a rich look and are shaped into squares/circular, hexagonal, rectangular, or even irregular beds both raised on the ground, if the area of garden exceeds 1 acre. These beds are essentially aligned with the building edge. In order to give a pleasing look, the length-to-width ratio of beds is preferred as 1:1.6. Taller shrubs and trees are planted in corners and away from walls, while climbers are placed near the entrances or along wall to give them support. Evergreens like Bougainvillea varieties—Mary palmer, Thimma, Shubhra, and Million dollar type-s—are the first choice as these remain in flower for 10 months in a year. Other such small shrubs giving continuous flowering like Russelia, Tabernemontana, Gulnar, Gultara, Kaner, and Peela kaner are grown singly in corners. Smaller fragrant shrubs are jasmines which are planted in front beds. Periphery of the beds is normally given a hedge of a variety of color, shape, height, and texture by combining different plants. Duranta is most preferred and kept smaller (30–40 cm height). Other hedges are 1 m to 2.5 m high and that are normally placed along building to gradually merge with

Fig. 8.2 *Borzicactus samaipatanus*

the lawns. These also act as divider screen, greening the walls, green curtains, and most importantly protect the lawns and ornamentals from hot winds as they act as windbreaks. Potted foliage plants add aesthetic value to any building and a choice can be made based on location of such pots. Trees in streets, avenues, roadside, roundabouts, and road crosses are essential for shade, fragrance, and color of the flower. Evergreen rather than deciduous trees have great value in ameliorating the hot environment and provide aesthetic look. Different categories of plants used in landscaping in arid lands in India are listed in the following paragraphs and photos of a few of these have been provided (Figs. 8.1, 8.2, 8.3, 8.4, 8.5, 8.6, 8.7, 8.8, 8.9, 8.10, 8.11, 8.12, 8.13, 8.14, 8.15, 8.16, 8.17, 8.18, 8.19, 8.20, 8.21, 8.22 and 8.23).

Fig. 8.3 *Cereus hexagonus*

8.5 Ornamental Trees

Acacia auriculiformis, *Azadirachta indica*, *Acacia concina*, *Bambusa arundinacea*, *Butea monosperma*, *Bombax ceiba*, *Cassia fistula*, *C. javanica*, *C. marginata*, *C. nodosa*, *C. renigera*, *C. siamea*, *Delonix regia*, *Ficus benghalensis*, *Phoenix sylvestris*, *Gliricidia sepium*, *Pongamia pinnata*, *Millingtonia hortensis*, *Acacia salicina*, *Peltophorum ferruginea*, *Eucalyptus* spp, *Tabubea argentea*, *Cercidium floridum*, *Tecomella undulata*, *Albizia lebbek*, *A. amara*, *Thespesia populnea*, *Melia azaderach*, *Terminalia arjuna*, *Terminalia catappa*, *Moringa oleifera*, *Albizia richardiana*, *Bauhinia racemosa*, *Dalbergia sissoo*, *Morus Alba*, *Ficus glomerata*, *Madhuca latifolia*, *Ficus infectoria*, *Ficus religiosa*, *Mitragyna parviflora*, *F. retusa*, *Acacia bevinosa*, *Cassia renigera*, *Delonix regia*, *Tamarix articulata* (Table 8.1).

Fig. 8.4 Bougainvillea variety—sensation

8.6 Ornamental Shrubs

Acalypha, *Bauhinia variegata*, *B. acuminata*, *Bougainvillea spectabilis*, *Barleria cristata*, *Barleria cuspidata*, *Carissa carandas*, *Callistemon*, *caesalpinia*, *Cestrum diurnum*, *C.nocturnum*, *Cassia auriculata*, *Cordia biosseri, Cathranthus roseus*, *Crinum asiaticum*—sukhdarshan, *Crotons*, *Caesalpinia pulcherrima*, *Cycas revoluta*, *Dombeya aliporensis*, *Duranta plumeric*, *Dombeya gagiana*, *Dichrostachys nutans*, *Dieffenbachia picta*, *Dracaena*, *Erythrina indica*, *E. cristagalli*, *E. subrosa*, *E. versipatela*, *Hamelia patens*, *Hibiscus rosa sinensis*, *Lawsonia alba*, *Lagerstroemia indica*, *Lantana camara*, *Lantana selloniana*, *Manihot esculenta*, *Jatropha curcas*, *J. multifida*, *Jatropha gossypifolia*, *J. coccinea*, *Jasminum sambac-Mogra*, *Murraya exotica*, *Murraya koenigi*, *Nerium oleander*, *Nyctanthes arbor-tristes*, *Punica granatum(Flora Plano)*, *Plumeria acutifolia*, *Plumeria rubra*, *Pseuderanthemum sp.*, *Russellia juncea*, *Rosa chinensis*, *Sanchezia nobilis*, *Sensiviera* spp., *Tabernaemontana coronaria*, *Thevetia nerifolia or peruviana*, *Tecoma stans*, *Thuja compacta Yucca* (Table 8.2).

Fig. 8.5 Bougainvillea variety—Gladys hapburn

8.7 Suitable Plants for Hedges

Clerodendrom phlomoides, *C. inerme*, *C. aculeatum*, *Barleria cristata*, *Bougainvillea spectabilis*, *Cassia auriculata*, *Dodonaea viscosa*, *Nerium oleander*, *Lawsonia inermis*, *Pithecellobium dulce*, *Prosopis juliflora*, *Parkinsonia aculeata*, *Tecoma stans*, *Vitex trifolia*, *Carissa carandas*, *Calligonum polygonoides*, *Dichrostachys cinerea*, *Barleria prionitis*, *Aloe vera*, *Duranta plumeria*, *Grewia tenax*, *Barleria Cristata*, *Jatropha curcass*, *Adhatoda vasica*, *Euphorbia caducifolia*, *Agave americana*, *Opuntia dillenii*, *Poinciana pulcherrima*, *Tabernemontana coronaria*, *Tamarix articulata*, *P. juliflora* (Table 8.3).

8.8 Creepers/Climbers

Argyreia nervosa, *Aristolochia grandiflora*, *Asparagus racemosus*, *Clitorea ternatea*, *Ipomea palmata*, *Ipomoea biloba*, *Jasminum auriculatum*, *Quisqualis indica*, *Antigonon leptopus*, *Cocculus pendulus*, *Bougainvillea glabra*, *B. peruviana*,

Fig. 8.6 Bougainvillea variety—Shubhra

B. spectabilis, Tinospora cordifolia, Bignonia grandiflora, Ephedra foliata, Jacquemontia violacea, Jasminum humile, J. auriantatum, J. officinale, J.multiflorum, Cryptostegia grandiflora, Merremia dissecta, M. tuberosa, Tecoma grandiflora.

8.9 Shade-Loving Plants for Indoors/Covered Areas Outside

Acalypha godseffiana, A. hispida, A. hamiltoniana, A. wilkesiana, A. torta, Adiantum sp., *Alocasia sanderiana, A. macrorrhiza, Codiaeum blumei, Dieffenbachia picta, D. daglance, Dracaena hookeriana, Ficus elastica, Gynura aurantiaca, Monstera deliciosa, Pandanus sanderi, Pichardea pacifica, Pilea mucosa, Pothos aureus, Scindapsus pictus, Sansevieria trifasciata, S. cylindrica, Sanchezia nobilis, Tradescantia bicolor, Tradescantia zebrina (variegated leaves), T. virginiana, Zebrina discolor,*

Fig. 8.7 Bougainvillea variety—Gloriosus

8.10 Bulbous Plants

Amayrllis stylosum, Crinum asiaticum, Canna indica, Cyperus alternifolius, Hymenocallis littoralis, Narcissus candida, Polianthes tuberosa, Zephyranthes candida, Z. rosea, Z. sulphurea, Mirabilis jalapa, Nerina curviflora, Haemanthus kalbreyeri.

Fig. 8.8 Bougainvillea variety—Joe

Fig. 8.9 *Mammillaria bocasana*

Fig. 8.10 *Opuntia ficus indica*

Fig. 8.11 *Pereskiopsis velutina* in pots placed to give a landscape view

Fig. 8.12 Landscape of cacti and succulents

Fig. 8.13 Bronze colored *Ipomoea* sp. as lawn edge

Fig. 8.14 *Thevetia peruviana* as lawn hedge plant

Fig. 8.15 A view of Machia Biological Park landscaped by keeping natives intact

Fig. 8.16 *Clerodendrum inerme* hedge on both sides of walkway in Samrat Ashoka Udyan in Jodhpur

8.11 Palms

Calyptrocalyx spicatus, *Livistona*, *Kentia belmoreana*, *Pinanga kuhlii*, *Pritchardia patula*, *P. grandis*, *Adiantum*, *Asplenium*, *Nephrolepis*, *Pteris*.

8.12 Bougainvilleas Varieties

Alok, Begum sikander, Cherry blossom, Joey, Mahara, Million dollar, Meera, Pixie, Purple price, Dwarf Gem, Blondis, Superba, Brilliant, Garnet Glory, Poultonii Special, Mrs. Fraser, Mrs. Delux Perry, Summer time, Padmi, Lady hope, Gladys hepburn, Mrs. Mentales, Ruaraka Purple prince, Scarlet glory, Spring festival, Mrs. R. B. Carrick, Magnifica, Versicolor, Profusion, Truley's special, Sensation, Splendens, Manohar Chandra variegated, Sonnet, Elizabeth, Dr. R. R. Pal, Glabra, Gopal, Golden Glow, Happiness, Lady-Mery-Bearing, Maharaja of Mysore, Mahatma Gandhi, Mary palmer, Orange Scepter, Purple star, Partha, Rosea fuschea, Subhra, Splendens, Thimma, Zakaria.

Fig. 8.17 Native trees of *Prosopis cineraria* are integrated into the landscape plantation in the Samrat Ashoka Udyan in Jodhpur

8.13 Cacti and Succulents

Borazicactus samipatans (Card.) Kimn., *Cerus hexagonus* (L.) Mill, *Cerus jamacaru* DC, *Chamaecereus silvestrii*, *Echinocactus grusonii*, *Echinocactus pectinatus* (Scheidw.) Engelm., *Echinocereus pentalophus*, *Echinocereus* sp., *Echinopsis multiplex* (Peiffer) Peiffer & Otto, *Ferocactus emory* (Engelmann) Orcutt, *Ferocactus wislinzeni*, *Gymnocalycium* sp., *Hylocereus triangular* (L.) Britton & Rose, *Hylocereus undatus* (Haworth) Britton & Rose, *Mammillaria camptotricha*, *Mammillaria bocasana* Poselg., *Mammillaria longimamma*, *Mammillaria schiedeana*, *Mammillaria variculeata*, *Myrelocactus geometrizans* (Mart.) Console, *Notocactus horstii*, *Opuntia cylindrica* DC, *Opuntia ficus indica* (L.) Mill., *Opuntia fulgida*, *Opuntia glomerata*, *Opuntia microdasys* var. *albispina*, *Opuntia microdasys* var. *pallida*, *Opuntia microdasys* var. *rufida*, *Opuntia salmiana*, *Opuntia vestita*, *Pereskiopsis velutina*, *Stapelia* sp., *Trichocereus pachanoi*, *Opuntia martolensis* Britt. & Rose, *Agave weberi* (Kumar and Kulloli 2016).

Fig. 8.18 Bougainvillea as boundary hedge is common in the desert cities in India

Fig. 8.19 Landscaping at lawn edge with *Cycas revoluta* row followed by taller *Nerium oleander* and then bigger trees in a public garden in Jodhpur

Table 8.1 Suitable ornamental trees for arid areas

S.no	Scientific name	Family	Local name	Flower colour	Flowering time	Canopy spread (m)	Plant height (m)	Suitability for locations*
1.	*Pongamia pinnata* Pierre	Papilionaceae	Pit-Papri	Pink	April	4–6	3–5	2, 3, 8, 9, 11, 12
2.	*Bauhinia alba* var *purpurea* L.	Caesalpinaceae	Kachnar	White Red	April–May	4–6	3–4	3, 6, 8, 9, 11, 12
3.	*Bombax ceiba* L.	Malvaceae	Simul	Bright Crimson	Jan–Feb	6–8	6–10	3, 7, 8, 9
4.	*Butea monosperma* (Lam.) Taub.	Leguminosae	Dhak, Palash	Salmon Red	Feb–March	6–8	6–8	3, 4, 5, 8, 9
5.	*C. marginata* Roxb.	Leguminosae	Red Cassia	Dull pink	July–Aug	6–8	7–8	3, 4, 5, 8, 10, 11, 12
6.	*C. nodosa* Buch-Han	Leguminosae	Pink Mohar	Pink, Deep Pink	May–June	5–7	5–7	3, 4, 5, 8, 10, 11, 12
7.	*C. siamea* Lam.	Leguminosae	Kala Sares	Yellow	June–Sept	6–9	6–9	3, 4, 5, 8, 10, 11, 12
8.	*C. surattensis* Burm. F.	Leguminosae	–	Yellow	Sept–Oct	5–7	5–6	3, 4, 5, 8, 10, 11, 12
9.	*Callistemon citrinus* (Curtis) Skeels	Myrtaceae	Bottle brush	Red, Yellow	March–April	3–4	5–6	1, 3, 4, 8, 10, 11, 12
10.	*Cassia fistula* L.	Caesalpiniaceae	Amaltas	Golden Yellow	April–May	5–7	6–8	3, 4, 8, 11, 12

(continued)

Table 8.1 (continued)

S.no	Scientific name	Family	Local name	Flower colour	Flowering time	Canopy spread (m)	Plant height (m)	Suitability for locations*
11.	*Cassia javanica* L.	Caesalpiniaceae	Pink Cassia	Red	June–May	6–8	5–8	3, 4, 5, 8, 10, 11, 12
12.	*Cupressus sempervirens* L.	Cupressaceae	Suru	_	_	4–6	15–20	1, 3, 4, 8, 10, 11, 12
13.	*Delonix elata* (L.) Gamble	Caesalpiniaceae	Safed Gulmohar	Yellowish white	Feb–March	6–8	8–10	3, 8, 10, 11
14.	*Delonix regia* Raf.	Caesalpiniaceae	Gulmohar	Orange–Red	April–May	5–7	6–8	3, 4, 5, 8, 11, 12
15.	*Erythrina indica* Lam.	Papilionaceae	Rakt Madar Pangara	Scarlet red	Feb–March	5–7	5–6	3, 5, 6, 7, 8, 10, 11
16.	*Eucalyptus camaldulensis* Dehnh.	Myrtaceae	Safeda	White Creamy	Nov–Dec	6–8	12–15	6, 7, 9
17.	*Gliricidia maculata* HBK	Fabaceae	Glaricida	Pink	Feb–Mar	4–6	4–6	1, 3, 4, 8, 10, 11
18.	*Hardwickia binata* Roxb.	Fabaceae	Anjan	Pale Yellowish	July–Sept	4–6	8–9	6, 7, 9
19.	*Melia azedarach* L.	Meliaceae	Bakaian	Lilac	June–July	5–7	9–12	1, 2, 3, 8, 10, 11
20.	*Millingtonia hortensis* L.	Bignoniaceae	Indian cork tree	White	Nov–Dec	5–6	10–12	2, 3, 4, 5, 7, 8, 10, 11

(continued)

Table 8.1 (continued)

S.no	Scientific name	Family	Local name	Flower colour	Flowering time	Canopy spread (m)	Plant height (m)	Suitability for locations*
21.	*Peltophorum ferrugineum* (DC.) Hayne	Leguminoceae	Pila Gulmohar	Rusty Yellow	May–Aug	8–10	10–12	1, 2, 3, 5, 8, 10, 11
22.	*Polyalthia longifolia* (Sonn.) Thwaites	Annonaceae	Ashokan	–	–	4–6	10–12	1, 2, 3, 4, 8, 10, 11, 12
23.	*Tabebuia argentea* (Bureau & K.Schum.) Britton	Bignoniaceae	Basant Raani	Yellow	March	4–5	6–8	3, 4, 8, 10, 11
24.	*Tecomella undulata* D. Don.	Bignoniaceae	Rohida	Orange Yellow	Nov–Feb	6–8	6–8	2, 3, 5, 8, 11
25.	*Acacia auriculiformis* A.Cunn. ex Benth.	Leguminoceae	–	Creamy Yellow	March–Jun	4–6	5–15	2, 3, 6, 7, 8, 9
26.	*Acacia salicina* Lindl.	Leguminosae		Yellow	Apr–Jun	4–6	5–8	2, 3, 5, 6, 8, 9, 11
27.	*Albizia lebbeck* (L.) Benth.	Leguminosae	Saresh	Creamy Yellow	May–Jun	12–18	15–20	2, 3, 6, 8, 11
28.	*Thespesia populnea* (L.) Sol. ex Correa	Leguminosae	Paras Pipal	Yellow to orange	Throughout year	6–8	8–12	1, 2, 3, 4, 5, 8, 11
29.	*Albizia amara* (Roxb.)B. Boivin	Leguminosae	Kaala Saresh	Yellow	May–Jun	3–5	5–8	2, 3, 6, 8, 9, 10, 11

(continued)

Table 8.1 (continued)

S.no	Scientific name	Family	Local name	Flower colour	Flowering time	Canopy spread (m)	Plant height (m)	Suitability for locations*
30.	*Terminalia arjuna* (Roxb. ex. DC.)Wight & Arn.	Combretaceae	Arjun	Creamy Yellow	Apr–Jun	8–10	10–15	3, 5, 7, 8, 9, 11
31.	*Terminalia catappa* L.	Combretaceae	Jungali Badam	Creamy Yellow	Mar–Jun	3–5	10–12	3, 5, 11, 12
32.	*Morus alba* L.	Moraceae	Shahtut	Creamy Yellow	Mar–Jun	4–6	5–6	3, 8, 11, 12
33.	*Ficus religiosa* L.	Moraceae	Peepal	Red Brown	Apr–Jun	15–20	15–20	2, 3, 4, 5, 7, 8, 9
34.	*Moringa oleifera* Lam.	Moringaceae	Sehnjana	White	Feb–Mar	5–6	5–6	3, 5, 7, 8, 9, 12
35.	*Ficus glomerata* L.	Moraceae	Gular	Red Brown	Apr–Jun	6–8	6–8	2, 3, 4, 5, 7, 8, 9
36.	*Dalbergia sisso* Roxb.	Leguminosae	Shisam	Light Yellow	Mar–Apr	4–6	5–6	1, 2, 3, 6, 7, 8, 10
37.	*Ficus infectoria* (Miq.) Miq.	Moraceae	Pakad	Cream colored	Jan–Mar	8–12	10–14	1, 2, 3, 4, 6, 8, 9, 11
38.	*Madhuca longifolia* (J. Konig.) J.F.Macbr.	Sapotaceae	Mahua	Creamy	Dec–Jul	10–12	12–15	2, 3, 5, 6, 7, 8, 9, 11
39.	*Mitragyna parviflora* (Roxb.) Korth.	Rubiaceae	Kadamb	Light Yellow	Mar–May	5–6	5–10	3,5,7,8,9
40.	*Ficua retusa* L.	Moraceae		Red Brown	Apr–Jun	4–6	8–10	3, 5, 8, 12

(continued)

Table 8.1 (continued)

S.no	Scientific name	Family	Local name	Flower colour	Flowering time	Canopy spread (m)	Plant height (m)	Suitability for locations*
41.	*Cassia renigera* Benth.	Caesalpiniaceae	Gulabi Cesia	Pink	May–Jun	4–6	6–8	3, 4, 8, 10, 11, 12
42.	*Tamarix articulata* Vahl.		Farash	Pink	Jun–Jul	4–6	8–10	3, 6, 7, 9, 11
43.	*Ficus benghalensis* L.	Moraceae	Bargad	Red	Mar–Apr	15–20	15–20	1, 2, 3, 4, 5, 7, 8, 9, 10, 11

*These numbers indicate locations on which this tree is most suitable for growing

1. Hospitals & Public dispensaries
2. Bus stand and Railway station including other public parking areas
3. Parks & Gardens visited by larger public, play grounds, stadia
4. Marriage places
5. Religious places Temples, Mosque, Gurdwara
6. Linear areas like highways, roads, railway line siding, avenues and streets
7. Water bodies like ponds, lakes, stream banks, canal banks
8. Schools, colleges, university campus, and places of education
9. Historical and ancient places like forts, palaces, or other memorials
10. Markets, bazaars, shopping complexes and malls
11. Institutional complexes
12. Home gardens

Table 8.2 Suitable ornamental shrubs for arid areas

S.no.	Scientific name	Family	Local name	Flower colour	Canopy spread (m)	Plant height (m)	Propagation method	Period of propagation
1	2	3	4	5	6	7	8	9
1.	*Acacia bavinosa* DC.	Fabaceae	–	Yellow	3–4	1–3	Seeds	July– September
2.	*Barleria cristata* L.	Acanthaceae	Kanteli	Yellow White, Blue	0.5–1.0	1–1.5	Cutting	July– September
3.	*Barleria prionitis* L.	Acanthaceae		Yellow	0.4– 1.5	1–2	Cutting	July– September
4.	*Bougainvillea glabra* Choisy	Nyctaginaceae	Boganbel	Red Yellow Mauve	4–5	4–5	Cutting	February, July–Aug
5.	*Bougainvillea spectabilis* Willd.	Nyctaginaceae	Boganbel	Red Yellow Mauve	5–6	4–5	Cutting	February, July–Aug
6.	*Cassia artimisioides* Gaudich ex. DC.	Fabaceae		Sulphur Yellow	0.5–1.0	1–2	Seed	Jul–Aug
7.	*Cassia auriculata* L.	Fabaceae	Anwal	Sulphur Yellow	2–3	2–2.5	Seed	Feb–Jul
8.	*Cestrum diurnum* L.	Fabaceae	Din ka Raja	Sulphur Yellow	1–2	2–2.5	Cutting	Jul
9.	*Cestrum nocturnum* L.	Solanaceae	Rat Ki Rani	Creamy white	1–2	2–3	Cutting	Jul

(continued)

Table 8.2 (continued)

S.no.	Scientific name	Family	Local name	Flower colour	Canopy spread (m)	Plant height (m)	Propagation method	Period of propagation
10.	*Clerodendrum inerme* (L.) Gaertn.	Verbenaceae	Venjai	White	2–3	2–2.5	Cutting layering	Feb–Jul
11.	*Clerodendrum aculeatum* (L.) Schltdl.	Verbenaceae	Jhurmutiya	White	1–2	1–1.5	Cutting layering	Jul–Oct
12.	*Dichrostachys cinerea* (L.) Wight & Arn.	Leguminoceae	Kalai	Pink Yellow	3–4	2–3	Seed	Jul
13.	*Dombeya viscosa* Hochr.	Malvaceae	Ittar donna	Pink Red	3–4	3–4	Cutting	Jan–Feb
14.	*Duranta plumeiri* Jacq.	Verbenaceae	Jhurmutta	Blue	0.5–1.0	2–3	Cutting	Jul
15.	*Euphorbia pulcherrima* Wild ex Klotz.	Euphorbiaceae	Patti	Scarlet Red	1–2	2–2.5	Cutting	Feb–Mar
16.	*Hamelia patens* Jacq.	Rubiaceae	Rat poison	Scrlet orange	2–3	2.5–3	Cutting	Jul
17.	*Hibiscus rosasinensis* L.	Malvaceae	Gudhal	Red, Pink	2–3	1.5–2.5	Cutting	Feb, Jul–Oct
18.	*Jasmina humile* L.	Oleaceae	Pili Chameli	Golden Yellow	2–2.5	3–4	Layering	Feb, Jul–Oct
1	2	3	4	5	6	7	8	9

(continued)

Table 8.2 (continued)

S.no.	Scientific name	Family	Local name	Flower colour	Canopy spread (m)	Plant height (m)	Propagation method	Period of propagation
19.	*Jasmine multiflorum* (Burm. f.) Andrews	Oleaceae	Dela	Milky white	2–3	2–3	Layering	Feb, Jul–Oct
20.	*Jasmine sambac* (L.) Aiton	Oleaceae	Mogra Motia	White	1–2	1–1.5	Cutting & Layering	Feb, Jul–Oct
21.	*Jasminum grandiflorum* L.	Oleaceae	Chameli	White	2–3	5–7	Cutting & Layering	Feb, Jul–Oct
22.	*Jasminum auriculatum* Vah.	Oleaceae	Joohi	White	1–2	5–7	Cutting, Layering	Jul–Aug
23.	*Jatropa acuminata* Jacq.	Euphorbiaceae	Jatropa	Scarlet Red	2–3	1.5–2	Cutting	Monsoon
24.	*Lagerstromia indica* L.	Lytheraceae		Red Pink, Mauve, white	3–4	3–5	Seeds	Jan–Feb
25.	*Lantana camara* L.	Verbenaceac	Tulsidas	White, Yellow, Mauve, Red	2–3	1.5–3	Seeds	July
26.	*Lawasonia inermis* L.	Lythraceae	Mehndi	Creamy White	2–3	2–3	Seeds, Cuttings	July–Aug
27.	*Murraya exotica* L.	Rutaceae	Kaminec	White	1–2	2–3	Seeds, Suckers	-do-

(continued)

Table 8.2 (continued)

S.no.	Scientific name	Family	Local name	Flower colour	Canopy spread (m)	Plant height (m)	Propagation method	Period of propagation
28.	*Neruim indicum* L.	Apocynaceae	Kaner	White Pink, Red	1–2	4–5	Seed, Cuttings	July
29.	*Nyctanthes arbortristis* L.	Verbenaceae	Har Singar	White	2–3	4–5	Seed	July
30.	*Parkinsonia aculeata* L.	Caesalpiniaceae	Jeru Salam Thron	Yellow	3–4	5–6	Seed	July
31.	*Plumeria acunitifolia*	Apocynaceae	Nagchampa	Yellow, White	4–6	3–5	Cutting	Feb–Jul
32.	*Plumeria rubra* Pit	Apocynaceae	Nagchampa	Red	4–6	3–5	Cutting	Feb–Jul
33.	*Poincetia pulcherrima* L.	Caesalpiniaceae	Gul Tura	Orange, Yellow	1–2	2–2.5	Seed	Jul–Mar
34.	*Punica granatum* L.	Punicaceae	Phool amar	Scarlet Red	2–4	2–3	Cutting	July
35.	*Russelia juncea* Zucc.	Sorophulariaceae	Corel Plant	Coral Red, Scarlet Red	1–2	1–1.5	Cutting, Layering	July
36.	*Taberneamontane coronaria* Wild	Apocynaaceae	Chandni	Milky	1–2	3–4	Cutting	July
37.	*Tecoma stans* H.D.&K.	Bignoniaceae	Pili	Golden Yellow	2–5	3–5	Seed	July
38.	*Thevetia peruviana* Merr.	Apocynaceae	Pili Kaner	-do-	4–5	3–4.5	Seed	July

Table 8.3 Suitable hedges for arid areas

S.no.	Scientific name	Family	Local name	Suitable height (m)	Time required to attain shape (months)	Period of planting	Method of planting	Cutting interval (days)
LOW								
1	*Agave americana* L.	Agavaceae	–	0.7–1.0	12	July–Oct	Root suckers, Bulbils	–
2	*Barleria cristata* L.	Acanthaceae	Kantcli	0.5–0.7	12–18	July	Seed	15
3	*Canna indica* L.	Cannaceae	Keli	0.5–0.7	12	July–Oct & Feb	Rhizome	90
4	*Hamelia patens* Jacq.	Rubiaceae	–	0.5–1.0	12	July	Rooted Cutting	30
5	*Vinca rosea* L.	Apocynaceae	Sadabahar	0.4–0.6	6	July–Oct & Feb	Seed/ Seedings	30
MEDIUM								
6	*Bougainvillea spectabilis* Willd.	Nyctaginaceae	Bougenbel	1–1.5	15–24	July,Oct, Feb,Mar	Rooted cuttings	30
7	*Calligonum polygonoides* L.	Polygonaceae	Phog	1–1.5	15–20	Jan & July	Rooted Cuttings	60
8	*Capparis decidua* (Forssk.) Edgew.	Capparaceae	Kair	1.5–2.0	24–36	July	Seed	60

(continued)

Table 8.3 (continued)

S.no.	Scientific name	Family	Local name	Suitable height (m)	Time required to attain shape (months)	Period of planting	Method of planting	Cutting interval (days)
9	*Cassia auriculata* L.	Caesalpiniaceae	Anwal	1–1.5	18–24	July	Seed	30
10	*Clerodendrum aculeatum* (L.) Schltdl.	Verbenaceae	Jhurmutta	1–1.5	18–24	July–Oct, Feb–March	Rooted cutting	30
11	*Clerodendrum inerme* (L.) Gaertn.	Verbenaceae	Jhurmutta	1–1.5	12–18	-do-	-do-	30
12	*Clerodendrum phlomoides* hort. ex DC.	Verbenaceae	Arni	1.5–2.0	12–18	July–Sept	-do-	30
13	*Dodonea viscosa*	Sapindaceae	Raellia	1–2	18–24	June–Sept	Seed	30
14	*Duranta plumieri* Jacq.	Verbenaceae	Jhurmuttia	0.8–1.0	18–24	July–Aug	Seed/Stem cuttings	30
15	*Euphorbia caducifolia* Haines	Euphorbiaceae	Thor	1.5–2	14–36	Feb & Oct.	Non Rooted Cutting	180
16	*Hibiscus rosa-sinensis* L.	Malvaceae	Gudhal	0.8–1.0	18–24	July–Oct	Rooted cuttings	60
17	*Lagerstromia indica* L.	Lythraceae	Sawni	1.5–2.0	18–24	July	Rooted Cutting/Seeds	30
18	*Lantana camara* L.	Verbenacea	Lantana	0.8–1.0	12–18	July–Oct	Rooted Cutting/Seeds	30

(continued)

Table 8.3 (continued)

S.no.	Scientific name	Family	Local name	Suitable height (m)	Time required to attain shape (months)	Period of planting	Method of planting	Cutting interval (days)
19	*Lawsonia inermis* L.	Lythraceae	Mehandi	1.5–2.0	18–24	July–Aug	Rooted Cutting/Seeds	30
20	*Lycuim barbarum* L.	Solanaceae	Murali	1–1.5	18–20	July	Rooted Cutting	60

Fig. 8.20 *Quisqualis indica* climber on the gate edge of a residential house, Jodhpur

8.14 Conclusion

Thus, a judicious selection of species, proper land preparation, and aesthetic planning of urban spaces result in attractive landscapes, which are pleasing to the eye and provide a healthy environment that refreshes one and all.

Fig. 8.21 A step public garden on a rocky plateau developed by overlayering 1-m-thick soil layer and planting hedges on both sides of stepped walkway in Jodhpur

Fig. 8.22 Hedge bordering a lawn in the roundabout in front of a building in Jodhpur

Fig. 8.23 *Polyalthia longifolia* tree has been topped to spread its canopy serving as shade for the car parked outside residence in Jaipur

References

Chadha KL, Choudhry BL (1986) Ornamental horticulture in India. ICAR, New Delhi, 217p
Chadha KL (ed) (2015) Handbook of horticulture. ICAR, New Delhi
Jindal SL (1970) Flowering shrubs of India. Publication Division, Govt. of India, New Delhi, 178p
Kumar S, Kulloli RN (2016) Cacti in desert botanical garden. ENVIS knowledge product. CAZRI, Jodhpur, 67p
Sabina GT (2009) Ornamental plants. NIPA, New Delhi, 324p
Trivedi PP (1987) Home gardening. ICAR, New Delhi, 328p

Part III
Local Food and Horticulture Education

Chapter 9
Urban Agriculture as a Tool for Horticultural Education and Youth Development

Mary A. Rogers

Abstract Youth in urban areas often lack access to natural areas, and thus a connection to plants and the environment. Garden-enhanced learning and place-based learning has experienced renewed interest since the early 1990s, as a way to reconnect youth with the natural environment and to improve health outcomes. School gardens are typically found in urban areas and used to enhance science-based learning outcomes. School gardens and summer and after-school garden programs in urban areas can also be used to improve attitudes and preferences for fruits and vegetables, potentially increasing consumption and leading to positive health outcomes. A comprehensive review of the literature shows that youth gardens can reduce stress, improve attitudes toward school, facilitate collaboration and teamwork and intercultural awareness, improve peer relations and prosocial behavior, and improve self-efficacy and self-esteem. Evaluations of garden programs indicate that involving youth in higher-order responsibilities such as garden planning and decision-making results in higher levels of participation and leadership development. Most urban garden programs are targeted toward elementary and middle school ages, and fewer examples exist in the literature on high school programs, despite the evidence that this cohort can benefit from urban garden-based programs. Literature reviews report a need for more rigorous quantitative evaluation of garden-based programs connecting programmatic activities to positive outcomes that can be used to maximize benefits and inform policy. School gardens tend to be concentrated in high-resource schools and more needs to be done to bring these experiences to underserved schools.

Keywords Garden-enhanced learning · School gardens · Green schoolyard Children's gardens · Environmental education · Experiential learning

---

M. A. Rogers (✉)
Sustainable & Organic Horticultural Food Production Systems,
Department of Horticultural Science, University of Minnesota, 305 Alderman Hall,
1970 Folwell Avenue, 55108 Saint Paul, MN, USA
e-mail: Roge0168@umn.edu

D. Nandwani (ed.), *Urban Horticulture*, Sustainable Development and Biodiversity 18, https://doi.org/10.1007/978-3-319-67017-1_9

9.1 Introduction

Urban environments, dominated by paved surfaces and hardscapes, leave few opportunities for youth to independently explore the natural environment. According to the most recent U.S. census, 63% of the U.S. population resides in cities (U.S. Census Bureau 2015). Cities typically lack natural habitats for exploration and urban parents are increasingly choosing highly supervised activities for their children due to safety concerns (Carver et al. 2008). Without designed green spaces and planned exposure to nature, urban youth are unlikely to connect with the natural environment in meaningful ways (Blair 2009; Finch 2004) and are less likely to value environmental sciences and less likely to exhibit environmentally responsible behaviors (Blair 2009; Vaske and Kobrin 2001). As a result, youth may exhibit what has been coined "nature deficit disorder" (Louv 2008). Urban agriculture is an effective way to reconnect people to nature and is increasingly used as a teaching tool in schools, particularly in urban areas where access to natural areas is limited and can alleviate some of the negative effects of poverty and nature deficit (Kuo 2001). Although urban school gardens were once commonly used in the U.S., they declined after the First World War, as federal funding dwindled and outreach programs began to shift focus toward production and home gardens (Trelstad 1997). Since the early 1990s, there has been renewed interest in garden-enhanced education. This may be attributed to the convergence of recent public awareness around childhood obesity, the "nature deficit disorder" and interest in place-based experiential education (Louv 2008; Williams and Dixon 2013). Summer and after-school programs in urban agriculture offer additional garden experiences for urban youth. Multiple benefits have been attributed to garden-enhanced learning, including improved academic performance and attitudes toward school, increased agricultural literacy, and food knowledge, improved health and nutrition outcomes, environmental awareness, emotional development and prosocial behaviors, and leadership and skills development (Blair 2009; Graham et al. 2005; Hirschi 2015; Ozer 2007; Robinson-O'Brien et al. 2007; Robinson 2016; Williams and Dixon 2013).

School gardens connect the garden experience to curriculum and there have been a number of programs and activities developed to support garden-based learning (Robinson 2016). Most school gardens address science-based outcomes, followed by language arts, math, writing, and social studies (Williams and Dixon 2013). There has been wide interest in using school gardens to promote positive attitudes toward fruits and vegetables among youth, likely due to the recent obesity epidemic. Public health and nutrition professionals predict that hands-on gardening activities combined with nutrition education will result in increased consumption of these foods (Lineberger and Zajicek 2000; McAleese and Rankin 2007; Morris and Zidenberg-Cherr 2002; Parmer et al. 2009; Radcliffe et al. 2011). Many of the nutritional and public health studies indicate positive effects, although more rigorous quantitative data is needed to conclusively connect garden activities with positive health outcomes and guide future research and policy. Youth gardens have also been used for environmental education and awareness and fostering appreciation for nature (Asah et al. 2012;

Skelly and Zajicek 1998). Garden activities can be designed to foster personal and emotional development and prosocial behaviors, as nature immersion reduces stress, can provide a safe haven and fosters cooperation and teamwork among youth, including those who may be perceived as different or "other" (Chawla et al. 2014; Cutter-Mackenzie 2009; Kim et al. 2014). In addition, increasing levels of responsibility and decision-making can empower youth and help develop leadership skills and lifelong learning (Eames-Sheavly et al. 2007; Hart 1997; Lekies et al. 2009).

9.2 History of School Gardens

Connecting garden experiences to youth education is not a new concept. The first urban school garden in the U.S. was established in 1891 at the George Putnam School in Roxbury, MA, sponsored by the Massachusetts Horticultural Society (Trelstad 1997). The Nature-Study movement championed the use of school gardens in early education. Liberty Hyde Bailey, renowned Dean of the College of Agriculture at Cornell University, was a leader of the Nature-Study movement, which sought to reinforce the connection between youth and the outdoors, promote agrarian values, and provide interactive educational experiences through the establishment of outdoor classrooms that allowed for science lessons and observations of natural phenomena. John Dewey, along with Maria Montessori, were also early champions of garden-enhanced learning, echoing the views of the Nature-Study movement and emphasizing the importance of school gardens to cultivate life skills and practical education (Hirschi 2015). It is estimated that there were 80,000 school gardens in the U.S. in 1910. (Robinson 2016). As time went on, the focus moved away from gardens as educational tools and more toward utilitarian ends. From 1900 to 1915, progressive reformers advocated for schoolyard gardens in urban areas to reduce juvenile delinquency, assimilate children of immigrants, build workforce skills for youth, improve health and nutrition, while reducing urban blight (Trelstad 1997). Widespread support for school gardens fizzled after WWI when federal funds for gardening efforts waned and urban city center populations shifted due to immigration quotas and the rising popularity of suburbs and increased focus on home gardens. Additionally, other summer recreational opportunities such as park and summer camps became more popular, decreasing the availability of youth to work in school gardens throughout the summer (Trelstad 1997). Today, garden-enhanced learning is experiencing a resurgence in popularity, with nearly 27% of public elementary schools in the U.S. using school gardens in 2012–2013 (Turner et al. 2014). The state of California initiated a "Garden in Every School" program in 2006, encouraging every school to integrate garden-enhanced learning into the curriculum (Robinson 2016). Education pioneers including Dewey and Montessori laid the foundation for what is known today as "place-based curriculum", a multidisciplinary and experiential approach to teaching that is community-oriented and "prepares people to live and work to sustain the places they inhabit and to participate actively in democracy" (Woodhouse and Knapp 2000).

9.3 Benefits Garden-Enhanced Learning

Typical gardens allow youth the opportunity to practice horticultural skills including sowing seeds, growing seedlings, irrigation, fertilization, planting, weeding, picking flowers, harvesting fruit and vegetables, and practice pest management. Multiple benefits have been ascribed to school gardens and garden programs and addressed in the literature. Youth garden programs are most commonly used to promote and improve: (1) science education and academic performance, (2) personal and social development, (3) environmental attitudes, (4) food literacy and healthy eating habits, (5) school bonding, parental involvement, and formation of community (Williams and Dixon 2013). Garden-enhanced learning commonly takes place in school gardens, but after-school and summer garden programs and community gardens are also used to engage youth. Although all children may benefit from garden-based programming, youth in urban areas are more likely to experience nature deficit disorder (Louv 2008). Youth in urban areas have been increasingly cut off from independently exploring their environments due to safety concerns (Carver et al. 2008; Gaster 1991). Increased traffic and physical hazards impose limitations, and youth in impoverished areas may not have access to safe play spaces due to mismanagement or deterioration of playground equipment and crime in parks and neighborhoods (Carver et al. 2008; Gaster 1991). As a result, urban youth spend more time indoors and in supervised activities with limited exposure to the natural environment. This can be detrimental, as access to green space and plants have been shown to reduce stress and mitigate negative effects of urban poverty (Kuo 2001).

The renewed interest in schoolyard gardens started in the early 1990s and is continuing today. A recent survey shows that most elementary school gardens are located in the western U.S., in urban areas, in larger schools, and focus on middle to higher socioeconomic status (Turner et al. 2014). Lower income youth lack the same access to school gardens, due to lack of resources and support. Successful school gardens require a high level of administrative support and parental and community involvement. The most effective programs are participatory and engage youth in decision-making and leadership, and allow them control over their own learning. Most garden-enhanced learning programs serve youth in grades K-6 (approx. 5–12 years old), with fewer programs geared toward middle school students and even fewer focused on pre-K and high school students. Many of the studies documenting benefits of garden-enhanced learning include a mix of qualitative and quantitative measures, although it is widely reported that rigor of quantitative studies remains problematic (Blair 2009; Ohly et al. 2016; Phibbs and Relf 2005; Robinson-O'Brien et al. 2007; Williams and Dixon 2013).

9.3.1 Science and Academic Achievement

Garden-enhanced education is reported to increase scientific and academic competencies, connection to nature, and promote positive attitudes toward school (Ozer 2007). School gardens require a tight connection between garden activities and academic standards. Few reports have focused on the use of school gardens in high schools and the majority of studies have included students in 2nd–6th grade (Williams and Dixon 2013). When high schools use garden-enhanced learning, they are typically linked to a few specific courses, such as agricultural studies, rather than general STEM-based curricula (Graham et al. 2005). Garden-based curricula are predominately targeted toward elementary students. Popular garden curricula include Texas A&M's Junior Master Gardener Program (Dirks and Orvis 2005); Growing Minds, a program of the Appalachian Sustainable Agriculture Project; the GrowLab curricula developed by the National Gardening Association; 4-H Youth garden programs; the National Wildlife Federation Schoolyard Habitat Program, and various state Departments of Education and Agriculture have developed programs (Blair 2009; California Department of Education 2002; Louisiana Department of Education 2004; Minnesota Department of Agriculture 2013; Ozer 2007; Williams and Dixon 2013).

School gardens have long been framed as "living laboratories" that allow children in urban areas to observe biological processes while also connecting with the natural world (Trelstad 1997). A growing number of children are raised in metropolitan areas and may have limited opportunities to interact with natural elements such as plants, soil, water, and trees, in the same way rural children do (Blair 2009). School gardens offer the opportunity for children to explore important scientific concepts such as growth and decay, insect–plant interactions, nutrient cycling, and soil ecology; as well as connect with the humanities through the inclusion of lessons on the cultural and historical aspects of agriculture, nature writing and literature, and how humans interact with the natural world (Blair 2009; Ozer 2007). School gardens also offer the opportunity for teaching food systems and understanding various parts of the system from "seed to plate". Hands-on gardening experiences can help students better grasp complex food systems concepts including organic agriculture, local versus imported foods, energy costs associated with production, garden inputs, externalities, and marketing and economics. A handful of studies have attempted to quantify the impact of garden-enhanced learning on scientific achievement in elementary schools. Williams and Dixon (2013) provide a review of the impact of garden-based learning on academic outcomes in schools, highlighting studies published between 1990 and 2010. Klemmer et al. (2006) looked at the effects of a school garden program on scientific literacy of 3rd, 4th, and 5th graders based on science test scores. The authors found that school garden experiences improved test scores significantly for 5th grade participants over the control group, with both female and male students benefitting from the gardening activities. Another study introduced a hands-on gardening curriculum [based on the Junior Master Gardener Handbook Level 1, (Whittlesey et al. 1999)] into three inner-city 5th grade classrooms in East Baton Rouge Parish, LA (Smith and Motsenbocker 2005). The researchers found that science achievement scores showed improvement in the experimental group post-intervention, however, there was no significant difference between the garden-enhanced learning group and the control group. This was attributed to the lack of training for the volunteers who were tasked with leading the gardening activities and the lack of continuity between the garden activities and normal classroom lessons. A greater degree of success might result in greater participation from experienced educators. However, educators are often challenged in adopting garden-enhanced learning such as teacher interest in gardening, time, experience with gardening, curricular materials linked to changing academic standards and expectations, teacher knowledge and training, funding, and administrative support (Blair 2009; Graham et al. 2005).

Nature immersion is known to provide psychological benefits, reduce stress, and improve interpersonal relationships (Chawla et al. 2014). Although not directly tied to academic performance, school gardens may improve attitudes toward school and consequently help improve student performance. Waliczek et al. (2001) evaluated whether integrating gardening activities improved student interpersonal relationships and attitude toward school. There were no statistical differences between treatment groups, however, female students had more positive attitudes toward school after the gardening program than male students, and older students (7th graders) had higher positive interpersonal relationship scores than younger students (Waliczek et al. 2001). The authors suggested that when students are provided with more independence and control, attitudes toward school improves. This study also showed that different groups of students may experience garden-based learning differently and demographics such as gender, age, ethnicity should be considered in analysis.

Summer garden programs for youth may enhance science skills and literacy even when science education is not a direct programmatic goal because biological concepts are imbedded in the gardening experience. A qualitative ethnographic study was conducted with inner-city youth in a summer gardening program and included twenty-three youth (14 years old) (Rahm 2002). The study showed that youth were able to learn science concepts in a garden program that did not have science education directly featured in their goals. The garden provided a place for emergent learning opportunities as youth participants were encouraged to ask questions of the adult Master Gardener volunteers. These questions were triggered by the gardening work, which led to further youth-initiated learning opportunities. This provides an example of the garden as a learning environment where youth can make their own meaning and allow for students to practice rather than learn about science as is typical in a formal classroom setting. Garden-enhanced learning can complement and reinforce scientific literacy by offering a place to explore ecological and food system complexity and cycles, addressing basic benchmarks for science literacy (American Association for the Advancement of Science 2009; Rahm 2002). Furthermore, urban garden programs provide a unique opportunity to integrate science and civic engagement, as students work with organizations on tasks that contribute to the betterment of the overall community.

In a review of the literature on garden-enhanced learning and academic outcomes, 93% of the studies (14 of 15 total) showed positive outcomes for science, and 80% of the outcomes in math, and 72% in language arts had positive outcomes. This provides strong support that integrating garden experiences into the classroom promotes scientific literacy and overall academic achievement. Furthermore, 5th graders with access to school gardens in Washington, D.C. had higher science scores in standardized tests than those in schools without access to gardens, controlling for race and socioeconomic status (Ray et al. 2016). Garden programs outside the school may increase science literacy and allow for youth-initiated learning and a rich out-of-school learning experience (Rahm 2002). Although teachers and administrators perceive academic benefits from school gardening (Graham and Zidenberg-Cherr 2005), very little research has been done substantiating an improvement in science achievement as a result of garden-enhanced learning. Challenges to conducting this research may include the time required for planning and conducting research, recruitment and retention of human subjects, communication problems among multiple program partners with different goals and motivations, and funding for this research (Phibbs and Relf 2005). Very little published research reports demographic information, such as gender, ethnicity, and socioeconomic status of the student subjects are available. This information would provide more detail on whether or not high-need students are benefiting from garden experiences and whether garden-enhanced learning can help close the achievement gap (Ray et al. 2016).

9.3.2 Nutrition and Health

A major impetus for implementing garden-based education is to achieve improved health outcomes for youth. Most children in the U.S. do not get their recommended daily servings of fruits and vegetables. According to the USDA, girls should eat 1–1.5 cups of fruit and 1.5–2.5 cups of vegetables daily, and boys should eat 1–2 cups of fruit and 1.5–3 cups of vegetables daily (United States Department of Agriculture Choose My Plate 2017). A 2015 study shows that total fruit consumption has increased 17% for children from between 2009 and 2014, however, vegetable consumption was down 2% over the same time period (Produce for Better Health Foundation 2015). Fruit and vegetable consumption for both adults and children in the U.S. is below recommended levels (Centers for Disease Control and Prevention 2016). Increased fruit and vegetable consumption promotes health and prevents cardiovascular disease, obesity, certain cancers, and reduces risk of stroke (Van Duyn and Pivonka 2000). Practitioners point to the importance of early intervention to establish healthy eating and fitness habits (DiNubile 1993, Heimendinger and Van Duyn 1995). Educators and health professionals have experimented with pairing garden activities with nutrition education to improve attitudes toward fruits and vegetables in efforts to increase consumption among children. Many studies evaluating the effectiveness of gardening and nutrition education show positive results grounded in social cognitive theory (SCT), which describes the interaction between environmental, personal, and behavioral factors to explain and predict changes in behavior (Bandura 1986; Reynolds et al. 1999). Based on SCT, increased knowledge of gardening and nutrition will increase confidence and self-efficacy, improving preference for fruits and vegetables and result in a change of behavior (i.e., increased consumption of fruits and vegetables) (Lineberger and Zajicek 2000; Morris and Zidenberg-Cherr 2002). Nutrition programs have been implemented with schoolyard garden activities as well as delivered to youth through summer and after school programs. Most of these intervention strategies have focused on elementary school (K-5), with few studies linking nutrition education and gardening to preschoolers and high school students.

Youth may benefit from hands-on garden-based nutrition programs in comparison to traditional classroom-based education, mostly by improving preference and attitudes toward fruits and vegetables. A nutrition education program paired with school garden activities for 2nd graders showed that participating students were better able to identify vegetables, scored vegetables higher in taste tests, chose more vegetables at lunch, and ate more vegetables than students who did not participate in gardening activities (Parmer et al. 2009). Using the SCT framework, this school garden program was shown to increase the ability to identify, willingness to try, and preference for vegetables among 6th grade students (11–13 years old) compared to the control group who did not participate in gardening activities (Ratcliffe et al. 2011). Similarly, a garden and nutrition program increased preferences for vegetables among three 4th grade classrooms (9–10 years old) from different schools in California, with gardening enhancing preference above classroom-based nutrition education alone; and showed nutrition knowledge was retained six months after the intervention, indicating long-term change (Morris and Zidenberg-Cherr 2002). Another study showed that a nutrition program paired with garden-based activities resulted in an increase in fruit and vegetable consumption among sixth graders (10–13 years old) in three classrooms in Idaho, based on self-reported food recall (McAleese and Rankin 2007).

Preschool, after school, and summer programs may provide additional opportunities to reach youth. One study (Heim et al. 2009) investigated a 12-week YMCA summer garden-based nutrition program provided to 4th, 5th, and 6th graders. According to surveys, participating children enjoyed learning about and taste testing fruit and vegetables and working in the garden. The intervention led to an increase in the number of fruits and vegetables eaten, vegetable preference, and asking behavior at home. Similar results were observed in a garden-based program involving pre-K children, ages 3–5 years old, who were enrolled in a Head Start preschool program in Texas (Sharma 2015). The Head Start program serves low-income families and the children who participated in the study were predominately African American, 28% of whom were overweight or obese. This study used the SCT framework and pre- and post-testing to show an increase in preference for fruits and vegetables in this cohort after the garden-based nutrition education intervention (Sharma 2015).

Although the SCT may improve attitudes, a change in attitude may not always translate to a change in behavior. A program named *Nutrition in the Garden* was designed to help integrate nutrition education into the classroom, and researchers evaluated whether the students developed more positive attitudes toward fruit and vegetables and whether this translated to increased consumption of fruit and vegetables from the garden (Lineberger and Zajicek 2000). The program involved 3rd and 5th grade students from five schools in Texas. Student attitudes toward vegetables improved after gardening, with female students responding more strongly than male.

However, fruit and vegetable consumption of students did not improve as a result of the garden program, based on 24-hour recall journals maintained by the students. Another study showed that there was no change in preference for vegetables nor an increase in nutrition knowledge after 4th grade students participated in an eight-week long after-school program emphasizing gardening and nutrition education (O'Brien and Shoemaker 2006). The authors indicated that gardening-enhanced nutrition education programs should carefully consider the length, timing, and particular aspects of the gardening experience to maximize program efficacy. A systematic review of school gardens on health and wellbeing of youth revealed that eight studies out of thirteen showed an increase in preference for fruits and vegetables among youth participants, however, most studies used subjective measures that are vulnerable to social desirability bias as children may be "eager to please" (Ohly et al. 2016). Furthermore, only two studies in thirteen resulted in significant increases in fruit and vegetable intake in youth participants, indicating that preference may not result in behavior change (Ohly et al. 2016).

The rise in childhood obesity and poor nutrition is most markedly suffered by low-income and minority children. Garden-enhanced education may improve nutritional quality in schools and summer programs and expose children to fruits and vegetables, increasing their attitudes and preferences for these foods and result in more children asking for these foods at home. Parents are generally responsive to dietary requests made by their children and children may be more influential than parents in deciding snacks and where to go out to eat (O'Dougherty et al. 2006). Targeting early ages is thought to improve nutrition and health at critical junctures of childhood growth and development (Byrd-Bredbenner et al. 1993) and early intervention may have a larger impact on behavior (DiNuble 1993). However, high school students require more daily servings of fruits and vegetables and are known to have the poorest school food environments based on availability of unhealthy foods (O'Toole et al. 2007; Story et al. 2009). Only 28% of high school students reported eating vegetables two or more times per day, and 6.7% reported eating no vegetables (Centers for Disease Control and Prevention 2016). Consequently, more garden-based interventions may benefit this cohort. Youth, particularly from low-income and minority populations, would benefit from repeated and consistent exposure to nutrition education and garden activities to promote positive attitudes toward fruits and vegetables and reinforce wellness messaging. It is important to note that although attitudes and preferences to healthy foods may be improved, access to healthy foods continues to be a challenge, particularly for residents of USDA designated "Food Deserts", areas where retail options for fresh fruits and vegetables are lacking (United States Department of Agriculture Food Access Research Atlas 2017). More rigorous quantitative studies are needed to validate the positive outcomes of garden-enhanced learning on health and wellbeing of youth.

9.3.3 Environmental Awareness

Garden-enhanced educational experiences for urban youth can result in increased awareness and positive attitudes toward the environment. Childhood experiences with nature, environmental education, and place-based outdoor activities can contribute to adult appreciation for nature and pro-environmental attitudes (Gifford and Nilsson 2014).

Evidence suggests that the more time children spend in nature-based activities, the more likely they are to seek out similar activities as adults (Asah et al. 2012). People who spend more time in natural surroundings are more likely to want to protect these areas. Children often lack access to nature in urban areas and urban agriculture may be the most practical way to facilitate a connection to nature and promote environmental awareness.

Skelly and Zajicek (1998) developed a garden program, *Project GREEN*, to help teachers integrate environmental education into the classroom. Results were obtained using the Children's Environmental Response Inventory and showed that 2nd and 4th grade students who participated in the program had more positive environmental attitudes than nonparticipants, with 2nd graders scoring higher than 4th graders. Student participation in outdoor activities results in better environmental attitude, and the impact may be greater for younger rather than older students. Very few longitudinal studies have been conducted to connect childhood garden experiences to attitudes and behaviors in adulthood. Lohr and Pearson-Mims (2005) surveyed 2000 adult residents of large metropolitan areas in the U.S. on their childhood experiences and exposure to gardens and nature. The surveys showed that active gardening as a child resulted in stronger positive attitudes toward trees and gardening as adults. Early experiences influenced attitudes that manifested in increased gardening behaviors as adults, such as attending gardening classes. Active childhood gardening, such as picking flowers, planting trees, picking vegetables, promoted adult appreciation for gardening and the environment. This study indicated that urban youth in nature deficient surroundings gain an appreciation for plants and nature through positive garden-based interventions.

9.3.4 Personal and Social Development

School gardens often focus on connecting experiential learning activities with academic goals, however, garden-enhanced learning is also used to foster self-efficacy, self-esteem, and prosocial and inclusive behaviors in youth (Blair 2009; Ozer 2007). Gardens and natural spaces have been shown to function as havens from stress and anxiety in school children from ages 6 to 18 (Chawla et al. 2014). In one study, ethnographic observations and student interviews showed that natural areas helped alleviate stress, improved focus, and self-efficacy as well as supporting positive social groups. Habitat observed included a wooded area for recess play in elementary school

(6–13 years old), outdoor classroom for older elementary school students (9–13 years old), and gardening programs for high school students (14–18 years old). One teen student viewed the garden as a sanctuary, reporting: "*It's almost like meditation, like my body is present but my mind just kind of drifts off and goes someplace else, and thinks about things … It's brainless tasks most of the time, so it's also like zenful, so you get to listen to things … I think about stuff, so I don't have to go home and think about it right before bed, so instead I can just go to sleep and stuff. I just feel happier in a way, and more at peace*" (Chawla et al. 2014). The connection between nature and mental health and wellbeing is well established in social science literature and horticultural therapy (Frumkin 2001; Louv 2008; Ulrich et al. 1991). Gardens can help youth who suffer from behavioral and emotional disorders including Attention Deficit Hyperactivity Disorder (ADHD) and anxiety and depression (Chawla et al. 2014; Chiumento 2012; Ohly et al. 2016). Urban and school gardens can provide regular access to nature as a safe haven for relaxation and reflection, but also promote prosocial behaviors.

An urban school gardening program in Korea (Kim et al. 2014) investigated the effects of gardening on peer relations of elementary school students (5th and 6th grades) via surveys and questionnaires administered to students. The authors looked at social factors including self-esteem, cooperation, self-understanding, expression, sense of achievement, sense of power, sense of belonging, intimacy, communication, responsibility, and reinforcement. The results showed that a gardening program consisting of 90 min per week for ten weeks resulted in difference in persistence of friendship and adaptability between friends, and a greater increase in individual peer status after the school gardening program. The program had a positive influence on peer relationships, sociality, and rule-following and collaboration were improved. This indicates that gardens may be a tool for promoting inclusivity and reducing ostracism among individuals who may be perceived as different or "other". The use of gardens as a tool to build inclusivity and social equity has not been fully explored in the literature (Ray et al. 2016). The Youth Farm Market Project is a gardening enterprise engaging urban youth (9–24 years old) in Minneapolis-St. Paul focused on community wellness, social justice, and youth empowerment. Youth participants in this program overall improved their general food skills including gardening, cooking, nutrition, and exhibited positive social behaviors such as respect for others and improved classroom behaviors (Lautenschlager and Smith 2007). In addition, participants exhibited improved willingness to try different ethnic and unfamiliar foods. Programs like this could be used as a model to build intercultural competency skills for youth. Additionally, gardens have been used to improve ESL teaching and create a safe and engaging space for immigrant children as well as help increase cultural awareness for teachers (Cutter-Mackenzie 2009).

9.3.5 Leadership Development and Skill Building

Gardening has been used to promote leadership and lifelong learning skills among youth, particularly when youth are engaged in higher levels of participation such as shared decision-making with adults. The "ladder of participation", developed by Arnstein (1969) is a useful tool to assess the depth of engagement with youth in community development work, using seven steps or levels of engagement (Eames-Sheavly et al. 2007; Hart 1992, 1997). The first three steps of the ladder consist of nonparticipatory actions including "manipulation", "decoration", and "tokenism", where youth have little to no input in project development, planning, and implementation (Hart 1997). The next four steps involve various levels of participatory actions between youth and adult leaders spanning "assigned but informed" to "youth-initiated, shared decisions with adults" (Eames-Sheavly et al. 2007; Hart 1997). An example of manipulation is to have youth plant an adult-designed and planned garden, but frame the project as if the children designed and built the garden themselves (Eames-Sheavly et al. 2007). A higher-order level of youth participation would be a project that is entirely youth-initiated and controlled, with the assistance of adults. Inviting youth to take leadership roles including planning, decision-making, directing activities, and organizing tasks and scheduling meetings can build skills and increase self-confidence, self-awareness, and enthusiasm (Lekies et al. 2009).

Using the ladder of participation framework, a three-year-long extension-research project titled *Greener Voices* aimed to improve youth participation in garden programs by increasing the level of higher-order involvement (Eames-Sheavly et al. 2007). This project included six sites in New York and Pennsylvania in rural, suburban, and urban settings, with youth 3–18 years old. The authors found that giving youth more leadership power improved youth engagement, enthusiasm for gardening, greater respect and interest, increased confidence, prosocial behaviors, and teamwork skills. Barriers to participation were adult perception of youth as incompetent and in need of close supervision to perform to expectation. Youth were also not accustomed to the new roles and power given to them, and older youth were better adapted to administrative tasks than younger children. Poor participation of youth in garden-enhanced programs is likely due to the failure of adults to involve them appropriately, and one way to identify age-appropriate activities is to involve the youth directly at the onset of the program (Driskell 2002; Eames-Sheavly et al. 2007).

The *Garden Mosaics* program developed at Cornell University and Cornell Cooperative Extension-NYC offers a model for youth participatory research (Krasny and Doyle 2002). In this program, youth (9–16 years old) and Extension educators took part in a Participatory Action Research design to co-develop interview questions and engage with the local community on an activity that would benefit the community. The overall goal was for youth to document ethnic gardening practices in urban community gardens. As a result of the program, the educators were able to expand their programs to urban audiences, the youth learned about gardening from garden elders and developed positive inter-generational relationships with them, and the garden elders received assistance in the garden and reported that they felt appreciated and respected, and enjoyed working with the youth (Krasny and Doyle 2002). This is an example of a project that can promote community cohesion, develop skills among youth, and result in horizontal knowledge exchange between multicultural community elders and University Extension.

The *Growing Minds* program developed at Texas A&M University Department of Horticultural Sciences involves curriculum designed for youth (3rd, 4th, and 5th grade) and includes horticulture, health, nutrition, environmental science, and leadership with hands-on garden experiences (Robinson and Zajicek 2005). Using a Youth Life Skills Inventory (YLSI), researchers were able to measure higher life skills in the garden-based treatment groups pre- vs. post-intervention. The YLSI included statements pertaining to teamwork, self-understanding, leadership, decision-making skills, communication skills, and volunteerism (Robinson and Zajicek 2005).

9.4 Case Studies

There has been an upswing in garden-enhanced learning opportunities for youth since the early 1990s. Today, one in four public elementary schools has a school garden. Elementary school gardens are used for a variety of purposes, including supplementing school meals, promoting healthy food choices and basic food skills, offering a place for experiential and project-based learning (fostering scientific literacy), and providing opportunities to connect to the broader community. Additionally, school gardens can help bridge social gaps and increase intercultural competency skills, and potentially narrowing the achievement gap between students from different backgrounds. The Bruce Vento Elementary School Garden in St. Paul, MN illustrates how the school garden is used to provide these services and also shows how gardens can meet the needs of students in multiple grade levels.

There are very few reports of school gardens in high schools, likely due to the focus on early intervention and challenges in integrating garden-based curricula with limited time and resources due to tight core academic standards and college prep activities. Adolescents from urban areas lack basic agricultural literacy, even when residing in agriculturally dominated states such as California (Hess and Trexler 2011). Teens have more control over their diets and most do not get their recommended daily servings of fruit and vegetables. Garden programs integrated into urban high schools could provide an opportunity to increase basic agricultural literacy and food skills into the curriculum, and may also provide business and marketing lessons that are valuable workforce skills. The Freeport High School Student Farm and Orchard provide an example of how a garden program can integrate valuable educational lessons while also providing practical work experience.

The Urban Garden outreach program at the MN Landscape Arboretum has developed over three decades. The program currently provides garden-based summer programs offering opportunities for youth to connect with nature, learn food and agricultural concepts, and develop leadership and lifelong learning skills. The *Growing Good* program developed by the MN Landscape Arboretum provides garden-based summer experiences customized for youth of multiple age levels, spanning early childhood to young adult. The program predominately serves young people in underprivileged urban neighborhoods in Minneapolis and St. Paul. Activities are designed appropriately to maximize youth engagement, with greater expectations for participation and leadership for the older youth in the *Growing to Lead* program.

9.4.1 Bruce Vento Elementary School Garden

by: Kirsten Saylor, School Garden Coordinator.

The Bruce Vento Elementary School Garden in St. Paul, MN started with the purpose to help feed families and enable students to learn how to grow their own. The garden was well supported by a number of staff, including the Lead Engineer and the Principal. By the second season, there was an appreciation that it needed to be better integrated into the curriculum in order to optimize the learning potential as well as the production potential. In addition, a theme of integration across the grades and benefit of having the whole school work in the garden emerged as a main benefit of the garden to date.

The school is two-thirds new American families that are 1st, 2nd or 3rd generation Hmong and Latino families and very recently immigrated Karen. The other third mainly consists mainly of African-Americans, with East African, Nepal, Caucasian families as well. Students come largely from the surrounding neighborhood, which has one of the lowest average income and highest poverty rate in St. Paul. Food, shelter, clothing are regular daily concerns for our families. We have translators for Karen, Hmong and Spanish and all written items are translated before they go out.

By the end of 2016–2017 school year we had made a number of improvements to the garden in order to (1) be a good learning space for students, (2) have necessary features for a school garden, and (3) put in place opportunities for families and community members to learn about and be welcomed into the garden.

The garden is located in the crook of the school building adjacent to the front entrance and sits up on a small hill and is visible to the rest of the community. There are six garden beds for food production, a permanent smaller bed for strawberries, a compost bin, and two pollinator gardens. Fall of 2016, the site was finally completely covered in wood chips and a large tool bin on site. In addition, there is water access right on site, and space within the school for long tools or other growing materials to be stored as needed. In the spring, signs were adding and decorative fencing was added to let people know where the pollinator gardens were and where not to walk.

The garden plan evolved in collaboration with the garden team and constant conversation with the Administration and individual teachers and grade teams. The goal was to give every student a chance to connect with the garden, if not plant in it.

First, we started with principle of "Planting It Forward". We turned the challenge of having the growing season the opposite of the school year into an opportunity to have students contribute to those that follow them. For instance, the crops that 4th grade plants in the spring will be harvested by the next set of 4th graders, and they will benefit from what the 5th graders planted for them. To make this work, each grade has its own set of plants so that teachers do not need to keep reinventing their curriculum, but rather deepen their understanding of a small set of plants—enabling them to link it to math, literacy, and social studies. The Garden Plan identifies some of the great learning opportunities around nutrition, growing them, harvest and how they reproduce with each of the plants each grade will plant. Most importantly, by knowing what they will plant each year, it is possible to engage community partners, such as SNAP-Ed or Food and Nutrition Services, in harvesting and cooking the crops in the fall—making it even more feasible for teachers to facilitate the full range of experiences associated with the garden. This setup also enables good crop rotation as each module was designed around a basic crop family, and the Garden Plan discusses the importance of rotating crops for preventing insect and disease and that different crops have different nutritional needs and benefits. In addition, Kindergarteners plant the seeds for sunflowers and other big bright flowers along the edge of the building. 2nd graders are responsible for the pollinator garden, as insects are part of their standards. Pre-K will take over the newly installed perennial herb bed, hopefully delighting in the textures, smells, and different shapes available.

The future of the garden is still uncertain as the School Garden Coordinator position was funded by a one-time grant through UMN Extension SNAP-Ed. There is still a need to have someone actually coordinate the garden as an outdoor learning space and ensure it as a good learning space. This year, we started the process of providing opportunities for student groups and social workers to adopt a smaller but important part of the garden, such as building trellises, taking care of the compost bin, planting bulbs around the school and fixing the garden beds and fencing. We also held a garden open house and started working on building a collaborative of nonprofit organizations to sustain weekly get-togethers in the garden for families to connect with one another, learn about gardening and other opportunities available to them. Our goal is to have the school garden become an asset to the neighborhood, and help grow relationships between people, so together they can connect to nature and the multiple gifts of the garden.

9.4.2 Freeport High School Student Farm and Orchard

by: Grant McCarty, Extension Educator, University of Illinois Extension.

Located in Northern Illinois in the Rockford area, the Freeport High School Student Farm and Orchard began in 2012. Monica Pierce, the Freeport High School Ag Teacher, oversees this program. Freeport High School has 61% low-income students and is racially diverse, with 56% of the student population listed as white, 23% black, 11% mixed race, and 9% Hispanic. Sophomore, junior, and senior students enrolled in the vegetable production course spend the spring semester choosing varieties, starting transplants in the greenhouse, and planting as the summer season starts. A 20 member CSA runs during the summer and is managed by Ms. Pierce. The farm employs 5–6 students in the summer that learn recordkeeping skills, customer service experience, and build upon their agricultural skills. In addition, a weekly market stand is available for clients that were unable to purchase a share. Marketing and business planning are two cornerstones of this program and are covered throughout the class and growing season. The Student Farm and Orchard is located near the high school and allows for tours, engagement with the community, and the use of the farm for other classes. Recent additions to the farm have included small fruits, a greenhouse, and an orchard.

Through these various crops, students are able to better connect with the food they are growing as out of 70 students in the agricultural program, only one lives on a farm. With a focus on sustainable agriculture, students are being prepared not only to go into agricultural careers but also acquire the confidence and skills that they need to grow their own food.

The program has led to renewed interest in local foods production within Freeport. Produce grown on site is used in local restaurants and excess donated to area food pantries. With a population around 25,000, Freeport has also seen high unemployment and loss of manufacturing/other industry jobs in recent years. The farm itself is also located within a food desert. The successes of Freeport High School's Student Farm has led to greater civic pride in the community as well as renewed industry in addressing urban agricultural issues. Future plans for the farm include day camps on local foods education for elementary students.

9.4.3 Growing Good at the MN Landscape Arboretum

by: Randy Gage, Learning Center Program Coordinator.

Growing Good is an Urban Garden outreach program of the University of Minnesota Landscape Arboretum working in partnership with youth-serving organizations in Minneapolis and St. Paul, MN. The program invites young gardeners to learn through hands-on horticulture in the garden and imbedded in the community. *Growing Good* is comprised of two programs that collectively engage urban youth from ages 5 to 25. The Children's Garden in Residence is a seven-week hands-on science-focused summer garden program for children aged 5–12 years old. *Growing Good* also consists of a program for urban teenagers include paid internships designed to expand horticulture knowledge through garden-based entrepreneurial and community engagement.

The children's Garden in residence uses the garden as a living laboratory for hands-on science learning. Growing Good partners with urban, youth-serving organizations who want to bring a summer hands-on science in the garden program to the young people they serve. Currently, the program partners with organizations in the Frogtown neighborhood of St. Paul, and the Phillips, and Near North neighborhoods of Minneapolis. Partnerships are governed by a partnership agreement that identifies the roles and responsibilities of the community partner and the Arboretum. Community partners typically provide garden space and the children for the program the Arboretum commits to the care and maintenance of the children's garden space and bringing a teaching team each week to deliver the hands-on, garden-based inquiry science curriculum. The Arboretum also agrees to raise money to support all costs associated with the program.

The curriculum used in the Children's Garden in Residence program has evolved over 30 years of program experience. The curriculum includes activities and approaches that utilize the garden as a living laboratory for hands-on science investigation. The curriculum is aligned with state science standards with an emphasis on practicing science process skills. In the summer of 2016, the curriculum was revised to include standards-based nutrition content that connects directly to the weekly garden snack. Planting seeds and tending to their growth over a seven-week summer program invites young learners to observe, predict, taste, wonder, measure, and experience change over time in the garden and gives them their very own place to grow.

As young people grow, their needs change. For many urban teenagers, a summer job is an important right of passage. They can begin to earn money while building crucial skills and exploring opportunities for their future. *Growing Good* paid internships are designed to use the garden as a tool to facilitate important adolescent skill development.

Starting as young as 12 years old, an urban teenager can apply for a *Growing Good* paid internship through one of the *Growing Good* community partner organizations. That first summer the intern will participate in *CityFresh Veggies*. The veggies interns participate in a business to business entrepreneurial model, where they grow specialty herbs and vegetables and sell the produce directly to restaurant chefs. Successful sales earn the interns a commission that is paid at the end of the summer. This entrepreneurial model encourages the interns to learn horticulture skills and practice key entrepreneurial skills such as preparing a product list, making a sales call, filling out an invoice, and building good customer service skills.

After a summer with *CityFresh* Veggies, an intern may apply to join a garden product development team. This team of interns takes a garden product idea, produces it and brings the product to market. In 2016, the new product team collaborated with local restaurant chefs to produce a cookbook that features some of the chef's favorite recipes that showcase *CityFresh* veggie produce. Another internship opportunity lies with the Pop-up Garden project. This intern team installs and maintains pop-up gardens at neighborhood nonprofits and small businesses. The interns learn about how the nonprofits serve the community, explore nonprofit job and career opportunities and collaborate to showcase growing healthy food in small urban spaces.

When young people turn sixteen, they have the opportunity to work directly for the MN Landscape Arboretum. *Growing to Lead* interns take on a leadership challenge. Last summer the team was hired by *Think Small* to help design and install a pop-up garden at home daycares in the city. The team installed gardens at 15 local daycares and developed and presented follow up lessons after the gardens had begun to grow.

As young people progress through the program they gain opportunity to build skills and leadership. Those that are particularly interested in the work can apply to be an apprentice and work directly with a *Growing Good* leader throughout the summer to learn and prepare for further program leadership. Many of the Children's

Gardens in Residence Instructors and *Growing Good* Internship facilitator positions are filled by young adults who have participated in the garden programs as young people.

The *Growing Good* paid internship program works closely with community partners to insure that the Youth Program Quality Assessment principles are applied at every level of programming. Insuring best practices in youth development programming is a key component to the *Growing Good* work with teenagers and young adults.

Working with young people in a garden setting is not without significant challenges. The work can be hot and dirty. It can be intimidating to make a sales call, or try to secure a business deal with a local entrepreneur. People learn by doing and being challenged to do new things. *Growing Good* helps to provide a safe place to try new things, be active in the community and in the garden. Finding meaningful connection to a young person's life is the central goal of the *Growing Good* program. Another significant challenge for the operation of the program is the reliance on partnership. *Growing Good* relies on partnerships with youth-serving organizations, entrepreneurs, donors, University personnel, and Arboretum support. The project is vital because of these partnerships and would not be possible without them.

The Arboretum's Urban Garden program has been in operation for over two decades serving hundreds of young people over the years, offering many opportunities to get their hands in the dirt for the first time. Some have continued with the program serving as instructors or facilitators as they entered their early adult years. Today, two members of the program leadership team are graduates of the program and have chosen college and career paths to strengthen and expand the impact of the *Growing Good* program for a new generation of young people.

9.5 Summary

School gardens have played a historic role in the education and socialization of youth in urban areas in the U.S. Within the last 30 years, the interest in garden-enhanced education has experienced renewed interest. Championed by parents, educators, public health professionals, and environmentalists, gardens have the opportunity to provide benefits in cognitive, social, behavioral, and psychomotor domains. Typically, school gardens are integrated into curriculum and used to teach general science concepts, environmental studies, nutrition (Graham et al. 2005), and language arts, math and writing (Willams and Dixon 2013). Educational gardens can be as minimal as potted plants or raised beds or take up a sizable area to supplement school food services. School gardens and after-school programs that include gardening experiences are also used to provide exposure to fresh fruits and vegetables in order to increase consumption and improve health outcomes. Other benefits of garden-enhanced learning include improved social and intercultural skills, improved self-esteem and school pride, and increased academic success in other areas (Ozer 2007). Gardens also provide youth an opportunity to reconnect with nature and reduce stress and anxiety (Chawla et al. 2014).

Despite the wide array of benefits ascribed to garden-enhanced learning in elementary and middle schools, there are few studies that examine this type of learning in high schools. One reason may be the emphasis on college preparation and the pressure for content-based instruction. When gardens are used in high schools, they typically are tied to elective agricultural curriculum rather than core general science education or skills-based learning (Graham et al. 2005). Another limitation often acknowledged in the literature is the lack of rigorous quantitative data to verify and validate benefits of garden-enhanced learning. These types of studies are difficult to perform. Problem areas cited include poor planning, communication problems with collaborators, and the amount of time and quality of collaboration, lack of funding to collect longitudinal data (Phibbs and Relf 2005). In addition, few studies include demographic information, such as gender, socioeconomic status, ethnicity, validity is rarely mentioned, and researcher bias may play a role in drawing conclusions (Williams and Dixon 2013). However, it is worth noting that the majority of both qualitative and quantitative studies document the positive value of garden-enhanced learning on youth education and development in one way or another. The literature also suggests that there is a disparity in access to garden-enhanced learning based on socioeconomics and race. The greatest barriers and challenges to school gardens include a lack of time, lack of funds or institutional support and physical spaces, a lack of teacher interest and knowledge in gardening, lack of experience in horticulture, lack of curricular materials linked to school gardening, and reliance on volunteers (Graham et al. 2005; Hirschi 2015). Efforts should be made to overcome these barriers and increase gardening and outdoor educational opportunities for all youth. In order to continue to provide gardens for the benefit of childhood education and influence policy, more rigorous studies are needed to quantify the benefits of school gardens.

References

Arnstein SR (1969) Eight rungs on the ladder of citizen participation. J Am Inst Plan 35:216–224

Asah ST, Bengston DN, Westphal LM (2012) The influence of childhood: operational pathways to adulthood participation in nature-based activities. Environ Behav 44:545–569

Bandura A (1986) Social foundations of thought & action: a social cognitive theory. Prentice Hall, Englewood Cliffs

Blair D (2009) The child in the garden: an evaluative review of the benefits of school gardening. J Environ Educ 40:15–38

Byrd-Bredbenner C, Marecic ML, Bernstein J (1993) Development of a nutrition education curriculum for head start children. J Nutr Educ 25:134–139

Carver A, Timperio A, Crawford D (2008) Playing it safe: the influence of neighbourhood safety on children's physical activity—a review. Health Place 14:217–227

Centers for Disease Control and Prevention (2016) Youth risk behavior surveillance—United States. MMWR 65:30–38. Retrieved 2 June 2017 https://www.cdc.gov/healthyyouth/data/yrbs/pdf/2015/ss6506_updated.pdf

Chawla L, Keena K, Pevec I, Stanley E (2014) Green schoolyards as havens from stress and resources for resilience in childhood and adolescence. Health Place 28:1–13
Chiumento A (2012) A haven of greenspace. Young Mind Mag 118:32–34
Cutter-Mackenzie A (2009) Multicultural school gardens: creating engaging garden spaces in learning about language, culture, and environment. Can J Environ Educ 14:122–135
DiNubile N (1993) Youth fitness—problems and solutions. Prevent Med 22:589–594
Dirks AE, Orvis K (2005) An evaluation of the junior master gardener program in third grade classrooms. HortTechnol 15:443–447
Driskell D (2002) Creating better cities with children and youth: a manual for participation. Earthscan Publ., UNESCO (United Nations Educational, Scientific, and Cultural Organization) Publ., Paris
Eames-Sheavly M, Lekies KS, MacDonald L, Wong KJ (2007) Greener Voices: an exploration of adult perceptions of participation of children and youth in gardening planning, design, and implementation. HortTechnol 17:247–253
Finch K (2004) Extinction of experience: a challenge to nature centers? (Or, how do you make a conservationist?) Directions: J Assoc Nature Admin, Special Issue, pp. 1–7
Frumkin H (2001) Beyond toxicity human health and the natural environment. Am J Prev Med 20:234–240
Gaster S (1991) Urban children's access to their neighborhood changes over three generations. Environ Behav 23:70–85
Gifford R, Nilsson A (2014) Personal and social factors that influence pro-environmental concern and behaviour: a review. Intl J Psychol 49:141–157
Graham H, Zidenberg-Cherr S (2005) California teachers perceive school gardens as an effective nutritional tool to promote healthful eating habits. J Am Diet Assoc 105:1797–1800
Graham H, Beall DL, Lussier M, McLaughlin P, Zidenberg-Cherr S (2005) Use of school gardens in academic instruction. J Nutr Educ Behav 37:147–151
Hart R (1992) Children's participation: from tokenism to citizenship. International Child Development Centre, United Nations Children's Fund, Florence
Hart R (1997) Children's participation: The theory and practice of involving young citizens in community development and environmental care. United Nations Children's Fund, Earthscan, New York, London
Heim S, Stang J, Ireland M (2009) A garden pilot project enhances fruit and vegetable consumption among children. J Am Diet Assoc 109:1220–1226
Heimendinger J, Van Duyn MAS (1995) Dietary behavior change: the challenge of recasting the role of fruits and vegetables in the American diet. Am J Clin Nutr 61:1397–1401
Hess A, Trexler CJ (2011) A qualitative study of agricultural literacy in urban youth: understanding for democratic participation in renewing the agri–food system. J Agric Educ 52:151–162
Hirschi JS (2015) Ripe for change: garden-based learning in schools. Harvard Education Press, Cambridge, MA
Kim S-S, Park SA, Son KC (2014) Improving peer relations of elementary school students through a school gardening program. Hort Technol 24:181–187
Klemmer CD, Waliczek TM, Zajicek JM (2006) Growing minds: the effect of a school gardening program on the science achievement of elementary students. HortTechnol 15:448–454
Krasny M, Doyle R (2002) Participatory approaches to program development and engaging youth in research: The case of an inter-generational urban community gardening program. J Ext 40:1–21
Kuo FE (2001) Coping with poverty: impacts of environment and attention in the inner city. Environ Behav 33:5–34
Lautenschlager L, Smith C (2007) Beliefs, knowledge, and values held by inner-city youth about gardening, nutrition, and cooking. Agric Hum Values 24:245–258
Lekies KS, Baker B, Baldini J (2009) Assessing participation in youth community action projects: opportunities and barriers. Community Dev 40:346–358
Lineberger SE, Zajicek JM (2000) School gardens: can a hands-on teaching tool affect students' attitudes and behaviors regarding fruit and vegetables? HortTechno 10:593–597

Lohr VI, Pearson-Mims CH (2005) Children's active and passive interactions with plants influence their attitudes and actions toward trees and gardening as adults. HortTechnol 15:472–476
Louv R (2008) Last child in the woods: saving our children from nature deficit disorder. Algonquin Books, Chapel Hill
McAleese JD, Rankin LL (2007) Garden-based nutrition education affects fruit and vegetable consumption in sixth-grade adolescents. J Am Diet Assoc 107:662–665
Morris JL, Zidenberg-Cherr S (2002) Garden-enhanced nutrition curriculum improves fourth-grade school children's knowledge of nutrition and preferences for some vegetables. J Am Diet Assoc 102:91–93
O'Brien SA, Shoemaker CA (2006) An after-school gardening club to promote fruit and vegetable consumption among fourth grade students: the assessment of social cognitive theory constructs. HortTechnol 16:24–29
O'Dougherty M, Story M, Stang J (2006) Observations of parent-child co-shoppers in supermarkets: children's involvement in food selections, parental yielding, and refusal strategies. J Nutr Educ Behav 38:183–188
O'Toole TP, Anderson S, Miller C, Guthrie J (2007) Nutrition services and foods and beverages available at school: results from the school health policies and programs study. J Sch Health 77:500–521
Ohly H, Gentry S, Wigglesworth R, Bethel A, Lovell R, Garside R (2016) A systematic review of the health and well-being impacts of school gardening: synthesis of quantitative and qualitative evidence. BMC Public Health 16:1–36
Ozer E (2007) The effects of school gardens on students and schools: conceptualization and considerations for maximizing healthy development. Health Educ Behav 34:846–863
Parmer SM, Salisbury-Glennon J, Shannon D, Struempler B (2009) School gardens: an experiential learning approach for a nutrition education program to increase fruit and vegetable knowledge, preference, and consumption among second-grade students. J Nutr Educ Behav 41:212–217
Phibbs EJ, Relf D (2005) Improving research on youth gardening. HortTechnol 15:425–428
Produce for Better Health Foundation (2015) State of the plate: 2015 study on America's consumption of fruit and vegetables. Retrieved 2 June 2017 http://www.pbhfoundation.org/pdfs/about/res/pbh_res/State_of_the_Plate_2015_WEB_Bookmarked.pdf
Rahm J (2002) Emergent learning opportunities in an inner-city youth gardening program. J Res Sci Teach 39:164–184
Ratcliffe MM, Merrigan KA, Rogers BL, Goldberg JP (2011) The effects of school garden experiences on middle school–aged students' knowledge, attitudes, and behaviors associated with vegetable consumption. J Health Promot 12:36–43
Ray R, Fisher DR, Fisher-Maltese C (2016) School gardens in the city: does environmental equity help close the achievement gap? DuBois Rev 13:379–395
Reynolds KD, Hinton AW, Shewchuck RM, Hickey CA (1999) Social cognitive model of fruit and vegetable consumption in elementary school children. J Nutr Educ 31:23–30
Robinson CW, Zajicek JM (2005) Growing minds: the effects of a one-year school garden program on six constructs of life skills of elementary school children. HortTechnol 15:453–457
Robinson-O'Brien R, Story M, Heim S (2007) Impact of garden-based youth nutrition intervention programs: a review. J Am Diet Assoc 109:273–280
Robinson CW (2016) Children and nature. In: Waliczek TM, Zajicek JM (eds) Urban horticulture. CRC Press, Boca Raton, pp 19–59
Sharma SV (2015) Feasibility and acceptability of a gardening-based nutrition education program in preschoolers from low-income, minority populations. J Child Res 13:93–110
Skelly SM, Zajicek JM (1998) The effect of an interdisciplinary garden program on the environmental attitudes of elementary school students. HortTechnol 8:579–583
Smith LL, Motsenbocker CE (2005) Impact of hands-on science through school gardening in Louisiana public elementary schools. HortTechnol 15:439–443
Story M, Nanney MS, Schwartz MB (2009) Schools and obesity prevention: creating school environments and policies to promote healthy eating and physical activity. Milbank Q 87:71–100

Trelstad B (1997) Little machines in their gardens: a history of school gardens in America, 1891 to 1920. Landscape J 16:161–173

Turner L, Sandoval A, Chaloupka FJ (2014) School garden programs are on the rise in US public elementary schools, but are less common in schools with economically disadvantaged student populations—a BTG research brief. Bridging the Gap Program, Health Policy Center, Institute for Health Research and Policy, University of Illinois, Chicago

Ulrich RS, Simons RF, Losito BD, Fiorito E, Miles MA, Zelson M (1991) Stress recovery during exposure to natural and urban environments. J Environ Psychol 11:201–230

U.S. Census Bureau (2015) U.S. cities are home to 62.7 percent of the U.S. population, but comprise just 3.5 percent of land area. Release Number: CB15–33. Retrieved 2 June 2017 https://www.census.gov/newsroom/press-releases/2015/cb15-33.html

U.S. Department of Agriculture (2017) Choose my plate. Retrieved 2 June 2017 https://www.choosemyplate.gov/dietary-guidelines

U.S. Department of Agriculture Economic Research Service (2017) Food access research Atlas. Retrieved June 2017 https://www.ers.usda.gov/data/fooddesert/

Van Duyn MAS, Pivonka E (2000) Overview of the health benefits of fruit and vegetable consumption for the dietetics professional: selected literature. J Am Diet Assoc 100:1511–1521

Vaske JJ, Kobrin CK (2001) Place attachment and environmental responsible behavior. J Environ Educ 32:16–21

Waliczek TM, Bradley JC, Zajicek JM (2001) The effect of school gardens on children's interpersonal relationships and attitudes toward school. HortTechnol 11:466–468

Whittlesey L, Seagraves R, Welsh D, Hall G (1999) Junior master gardener teacher/leaders guide. (Level 1. Texas A&M Univ. Agr. Ext. Serv., College Station)

Williams DR, Dixon PS (2013) Impact of garden-based learning on academic outcomes in schools: synthesis of research between 1990 and 2010. Rev Educ Res 83:211–235

Woodhouse JL, Knapp CE (2000) Place-based curriculum and instruction: outdoor and environmental education approaches (ED448012). ERIC Clearinghouse on Rural Education and Small Schools, Charleston

Chapter 10
The Role of Urban Horticulture in the Sustainable Agri-Food Movement

John Ikerd

Abstract The growing popularity of urban horticulture is being driven by growth in urban agriculture in general. This is particularly prominent in urban inner cities areas sometimes characterized as "food deserts." The motives for the urban agriculture movement in these areas are multifaceted. Critics who dismiss the movement for its lack of economic potential fail to understand that its social and cultural contributions are perhaps even more important than its significant contributions to intercity employment and food security. The urban agriculture movement is not new. Prior to the industrialization of agriculture, people in cities met many of their food needs from urban gardens and truck farms in peri-urban areas. The Garden City Movement of the late 1800s was a response to concerns for urban food insecurity that had resulted from reliance on the market-driven, industrial food system. Industrialization of agriculture, although successful in increasing production, had failed to provide domestic food security. Urban agricultural movements of the past have blossomed during times of food scarcity and income insecurity, particularly during times of war and economic recession. The current urban agriculture movement is different, in which it is continuing during times of agricultural abundance and economic sufficiency. Urban horticulture is more adaptable to urban areas than livestock or grain production. Thus, the current movement is commonly identified as an urban horticulture movement. The potential importance of the urban horticulture movement becomes readily apparent when it is viewed as part of the much larger sustainable agriculture or agri-food movement. Agricultural sustainability is not simply a matter of

[1]Food and Agriculture Organization of the United Nations, "Urban Agriculture," http://www.fao.org/urban-agriculture/en/.

[2]World Watch Institute, *State of the World 2011*, "State of the World Brief Series, Chapter 10, Feeding the Cities, http://blogs.worldwatch.org/nourishingtheplanet/wp-content/uploads/2011/02/Chapter-10-Policy-Brief_new.pdf?cda6c1.

J. Ikerd (✉)
Agricultural and Applied Economics, Division of Applied Social Sciences, University of Missouri, Columbia, MO 65210, USA
e-mail: jeikerd@gmail.com

D. Nandwani (ed.), *Urban Horticulture*, Sustainable Development and Biodiversity 18, https://doi.org/10.1007/978-3-319-67017-1_10

economic efficiency but also requires ecological, social, and economic integrity. Social and environmental benefits, which are prominent attributes of urban horticulture, are critical to agri-food sustainability. Lacking economic incentives, government policies will be essential to ensure the social and ecological benefits essential for agri-food sustainability. Past government policies supporting industrial agriculture have been justified as essential for food security, but have failed. The first condition of agricultural sustainability is to meet the basic food needs of the present—food security. The second requisite is to do so without diminishing opportunity for future generations. Government policies of the future should logically be redirected to ensure agri-food sustainability, including sustainable urban horticulture. Perhaps most important, the urban agriculture movement is empowering people in inner cities to take control of their own destinies. The urban agriculture movement exemplifies the ethics of social responsibility essential for agricultural sustainability.

Keywords Urban horticulture · Urban agriculture · Food deserts · Urban employment · Food security · Industrial agriculture · Urban gardens · Truck farms · Garden City Movement · Market driven food system · Food scarcity · Sustainable agriculture · Social benefit · Cultural benefits · Sustainable agri-food-government policies · Environmental benefits · Social benefits · Urban empowerment

The growing popularity of urban *horticulture* is being driven by a growth in urban *agriculture* in general. Urban agriculture and horticulture could well coevolve to become an important part of the food system of the United States (U.S.), as it now is in much of the rest of the world. The Food and Agricultural Organization of the United Nations estimates that more than 800 million people worldwide cultivate fruits and vegetables or grow livestock in cities.[1] The World Watch Institute estimates that urban agriculture produces 15 to 20% of the world's food.[2] The U.S. Department of Agriculture (USDA) does not yet keep statistics on urban agriculture in the U.S., but various studies indicate that urban agriculture is becoming an increasingly important source of fresh vegetables and fruits, particularly in economically decaying inner cities such as Detroit, MI, Philadelphia, PA, and Camden, NJ Royte (2017). The USDA has responded to increasing demands for information by providing new educational resources to help people producing foods in urban neighborhoods.

Motives for Urban Horticulture. Some critics see the trend toward urban agriculture as a passing fad brought on by public attention to urban "food deserts"—meaning urban communities without reasonable access to fresh fruits and vegetables. Critics contend that food production has moved away from cites for sound economic reasons and is not likely to return. Industrial agricultural technologies led to the geographic specialization of large-scale agricultural operations as an effective means of reducing costs of food production. Most critics do recognize that personal participation in food production provides individual intangible benefits and that community gardening has social benefits. However, they see little prospects for success of the urban agriculture movement in the face of continued economic globalization of food production.

A recent critique of urban agriculture in the *Journal of Agriculture, Food Systems, and Community Development* concluded: "It would be misleading to pretend that urban gardening could significantly improve food security and affordability Hallsworth and Wong (2013)." A set of rebuttal comments to the article focused on the importance of noneconomic motivations. One article focused on the "potential" for increased productivity in urban areas through utilization of existing and new technologies. Urban agriculture also provides the potential for fresher and more nutritious food Colasanti and Hamm (2013). Another rebuttal cited numerous studies showing the positive impacts of urban gardens on overall quality of life for people living in inner cities Lavid (2013). A third comment pointed out that many cities still have significant land available for urban gardening and that city governments have logical incentives to provide economic support to use this land for urban gardens Evans and Miewald (2013). The last rebuttal focused on the social mission of urban agriculture. "The growing movement is not predicated on false hopes of its productive potential, but recognizes urban cultivation as one of many approaches to address inequalities in the conventional food system Weissman (2013)."

Few, if any, of its defenders, claim that urban agriculture could create food self-sufficient cities. A Michigan State University Extension report summarizes the defense of urban agriculture as follows: "The practice of urban agriculture and community gardening as an organizational strategy, a comprehensive approach to supporting and engaging individuals and neighborhoods, promoting economic development through food production, improving neighborhood safety through combating blight, and exposing youth to intergenerational interactions and encouraging self-reliance has been a trend in recent years in the United States, particularly in Rust Belt cities that have experienced many years of economic decline McLean (2017)." A comprehensive review of literature, providing both pros and cons of urban agriculture, has been complied by the Johns Hopkins Center for Livable Futures.[3]

Urban Gardening: A Response to Agricultural Industrialization. Urban gardening certainly is not a new phenomenon. The early rural to urban migration was a consequence of economic industrialization. Large number of people were needed to work in the large industrial factories that were replacing craft-based economies. With limited transportation options, people needed to live near their places of work, so cites emerged and grew around factories. During much of the nineteenth century, many urban dwellers continued to produce much of their own food. To serve the food needs of others, "truck farmers" settled on urban fringes, trucking their fresh vegetables, fruits, milk, meat, and eggs into city neighborhoods for home delivery or sale at city markets.

As the cities continued to grow and became more densely populated they expanded into the countryside, replacing peri-urban farms with residential developments. Early city planners showed little concern for preserving spaces to produce food either within

[3]Raychel Santo Anne Palmer Brent Kim, "Vacant Lots To Vibrant Plots, A Review Of The Benefits And Limitations Of Urban Agriculture," file:///C:/Users/John/Documents/Articles-Abstracts/Urban%20agriculture/urban-ag-literature-review.pdf.

or near urbanizing areas. Green spaces within cities were largely reserved for parks with shade trees and scenic lakes, hillsides, and valleys, where people could escape the indignities of factory towns, not spaces where people could grow food. Larger plots of land in urban fringes became housing developments or residential estates. By the late 1800s, however, a few progressive city planners were becoming concerned about the dwindling opportunities to produce food in or near cities.

In 1898, Sir Ebenezer Howard initiated the "Garden City Movement" in Great Britain. His basic idea was to develop cities of modest size, ideally around 32,000 people, with the inner core of industry surrounded by residences and an outer green belt reserved for farms to provide food for the city.[4] In 1902, Howard published his classic book, *Garden Cities of Tomorrow* Howard (1902), which expresses his concerns about uncontrolled urbanization and the implications for the food security of those living in cities. However, his ideas were never widely accepted, and by the 1930s, only two model cities, Letchworth Garden City and Welwyn Garden City, remained in Great Britain.

Lewis Mumford of New York, a noted scholar, prolific writer, and believer in rational and ecologically sound urban planning, reframed Howard's concept of garden cities and succeeded in sparking renewed interest in urban agriculture Wojtowicz (2001). He began writing about urban agriculture in the 1920s, but his ideas became more popular in the 1930s. The Great Depression threatened the food security of cities with little space left to grow to food and made Mumford's ideas more appealing. His concept of urban agriculture is most notably expressed in, *The City in History, Its Origins, Its Transformations, and Its Prospects* Mumford (1961), which was published in 1961.

More generally, interest in urban agriculture has been a reoccurring phenomenon in modern American history. Past occurrences of growing popularity have coincided with crises that have led to widespread food scarcity. The popularity of urban agriculture surged during World War I and again during World War II when *Victory Gardens* and home gardening accounted for about half of total U.S. vegetable production.[5] Interest also surged during the economic depression of the 1890s, the Great Depression of the 1930s, and again with the economic recession of 2008. A 2001 United Nations report documents a historical tendency of urban agriculture globally to decline with industrialization and to resurge in popularity during times of local food scarcity Smit et al. (2001).

However, the current surge of interest in urban agriculture in the U.S. is notably different from those of previous cycles in that it is accelerating during a time when significant portions of agricultural production are being diverted to biofuel production and export markets. In addition, the vast majority of Americans have far more than enough income to buy enough food and meet their other basics economic necessities. Regardless, many Americans are left hungry, or at least at significant risk of not

---

[4]Geni, "The Garden City Movement," https://www.geni.com/projects/The-Garden-City-Movement/15255.

[5]Wikipedia "Victory Gardens," https://en.wikipedia.org/wiki/Victory_garden.

having enough food. Perhaps more significant, the foods affordable to many people in the inner cities contain more than enough calories but too few essential nutrients to sustain healthy, active lifestyles. The failure of the industrial food system to provide food security is both general and widespread. A larger percentage of Americans are threatened by hunger today than in the 1960s.[6] In 2013, more than 15% of U.S. residents were classified as "food insecure" and more than 20% of U.S. children were living in food-insecure households Coleman-Jensen et al. (2013). Food insecurity is simply more evident and most critical in cities where people have been left with limited access to fresh fruits and vegetables and little opportunity to grow their own food.

Urban Agriculture and the Quest for Sustainability. Ironically, the industrialization of American agriculture, which initially allowed agricultural to move out of urban areas, is now the primary motivation for returning agriculture to urban areas. The recent urban agriculture movement has coincided with the organic farming, sustainable agriculture, and local agri-food movements. All of these movements are rooted in a growing rejection of the industrial approach or paradigm of agri-food production and distribution.

The modern organic food movement emerged in the U.S. with the "back to the earth" movement during the 1960s. The organic food pioneers of the 1960s produced their own food, bought food from each other, and formed the first cooperative food buying clubs and natural food stores. Concerns about the health and environmental risks associated with the synthetic fertilizers and pesticides of industrial agriculture were important concerns but were not the only reasons the "back to the earth" communities chose to grow foods organically. They were creating and nurturing a sense of personal connectedness and commitment to taking care of each other and caring for the earth. The philosophy of organic farming was deeply embedded in their culture and their communities. They were pursuing an organic way of life.

Organics did not gain widespread public acceptance until the late 1980s, with the emergence of concerns for agricultural sustainability. Agricultural sustainability emerged out of growing concerns about the ability of industrial agriculture meet the basic responsibilities of agriculture to society. It not only failed to provide food security for consumers but also failed to provide quality employment opportunities for farmers and food industry workers, and it led to the social and economic decline and decay of both rural and urban communities. As industrial agriculture systematically pollutes and depletes the natural and human resources upon which its productivity and profitability ultimately depend it is diminishing the agricultural opportunities of future generations Ikerd (2008). By the 1990s, more people were beginning to understand that an industrial agriculture is not sustainable.

Sustainable agriculture is but one aspect of the larger sustainability movement that is challenging the industrial paradigm of economic development that now permeates virtually all aspects of so-called developed economies and societies. The industrial

---

[6]CBS documentary, "Hunger in America," 1968, https://www.youtube.com/watch?v=h94bq4JfMAA.

paradigm is characterizes by specialization, standardization, and consolidation of control into ever larger production units. Industrial organizations gain their initial economic advantages through the "economies of scale" of large-scale, specialized production. The initial economic efficiency is gained through *functional* specialization, or division of labor, which allows each worker to specialize in narrowly defined task or procedures in the production process. Functional simplification and routinization facilitate mechanization and consolidation of control into larger management units or organizations. The consolidation of industrial organizations required the concentration of large numbers of workers and the creation of cities. *Geographic* specialization or industrial organizations then facilitated more efficient exploitation of their economic comparative advantages associated with geographically-fixed natural resources—timber, minerals, and farmland.

Functional specialization of *agriculture* was made possible by post-World War II industrial agricultural technologies—notably affordable tractors and synthetic fertilizers and pesticides. Standardization of farming practices and simplification of farm management allowed farms to grow larger by employing fewer people. As farms grew larger and fewer, many farm families were forced to migrate to cities for employment. The geographic specialization of agriculture was made possible by industrial transportation technologies—notably refrigerated transport and interstate highways. Thus, the agricultural abandonment of cities during the latter 1900s was motivated and facilitated by the industrialization of American agriculture.

Eventually, more sophisticated industrial technologies—specifically, computers and robots—began displacing workers in industry as earlier technologies had done in agriculture. Factories were able to function more efficiently with fewer, less-skilled, lower paid workers. Manufacturing jobs then moved from urban to rural areas and then increasingly to other "less-developed" countries. Unemployed and underemployed workers in the old industrial cities were people left without the economic means to compete for food with growing market demand for biofuels and agricultural exports. The recent surge in of urban agriculture is motivated and facilitated by the failure of industrialization to provide *economic* security and the failure of the industrial food system to provide urban *food* security.

In the face of growing food insecurity, the urban agriculture movement is one dimension of a larger quest for agricultural sustainability. Sustainability, in the most basic sense, is the ability to meet the needs of the present without diminishing opportunities for the future. A sustainable food system then must be able to meet the "basic food needs" of current generations without diminishing the productivity of the natural and human resources needed to meet the basic food needs of future generations. Meeting the needs of all does not suggest that everyone must have everything they want, but instead must have everything they need to lead healthy, active, and purposeful lives.

Much of the focus of sustainability up to now had been on preserving opportunities for future generation through preservation and regeneration of natural resources. However, the first requisite for sustainability is to meet the needs of the present—meaning the basic needs of all. There are sound economically logical reasons for the inherent lack of ecological, social, or economic sustainability of an industrial food system.[7] However, the obvious failure of the industrial food system to provide enough nutritious food for all is sufficient motivation for those living in decaying inner cities to expect, or even demand, opportunities to grow their own food.

Attributes of Horticulture relevant to Urban Agriculture. Horticultural crops have several characteristics that make horticulture a particularly popular and valuable component of urban agriculture. For example, animal agriculture raises significantly more challenges in urban settings than do crops. Increasingly, cities are allowing chickens within city limits but typically with limits to numbers per household. Roosters are sometimes prohibited because their crowing can disturb neighbors. Chicken droppings or manure, although excellent fertilizer, may present problems for owners as well as neighbors if the birds are not restricted to enclosed areas.

Larger animals, such as hogs or dairy cows generally are restricted to less populated, peri-urban areas for reasons similar to those for restrictions on chickens in inner cities. Having not grown up around livestock or poultry, many urban dwellers simply are not comfortable living in close proximity to farm animals. Honey bees present similar challenges, such as neighbors' fear of bee stings. Animals also require more frequent and more personal attention than do crops. Farm animals cannot be left for extended periods of time without fresh water and periodic feeding and have no "dormant season" to give owners an extended break from farming.

This leaves fruits and vegetables as the crops of choice for most urban agriculturalists. Horticultural crops have specific advantages, in addition to their lack of disadvantages. Vegetables can yield more nutritional and economic value per acre, city lot, or square foot than can field crops such as corn, soybeans, or wheat. Even sweet corn, which is a popular food crop, requires more space per unit of production than tomatoes, salad greens, most root crops, or herbs. In addition to larger yields, fresh vegetables also tend to yield higher nutritional value than do most other crops suitable for the limited spaces in most urban areas. Lack of leafy green and yellow vegetables is often cited as the weak link in American diets. Freshly picked, flavor-filled vegetables tend to be scarce in inner-city "food deserts", and thus have significant economic value—even in low-income neighborhoods. However, the nutritional value of fresh vegetables to the families of urban gardeners may far outweigh their economic value in farmers markets.

Urban horticulture also requires less upfront cost or initial cash outlay than do most other urban agricultural options—such as food animals or aquaponics. The minimum requirement is access to a small plot of land, perhaps in the corner of a yard, or even a few clay pots on a patio. Requirements for equipment and supplies for a decent kitchen garden may be met with a spade, a hoe, and a few packets of seeds. In areas where the topsoil is gone or is toxic with chemical residues, a raised bed filled with good soil or compost may be required for decent yields. Regardless, it does not take a large outlay of cash to start a kitchen garden. Fruit trees require more

[7] Ikerd, *Crisis and Opportunity*, 2008.

space and upfront cash outlays, with delayed yields, but are nonetheless affordable to many urban residents.

The skill barriers to starting a small urban garden are also quite low. If the soil is fertile and well-drained, a bit of information regarding local planting dates and depths and water requirements will likely result in an edible harvest of crops adapted to the local growing region. Another essential skill is to be able to distinguish between crops and weeds and between insect pests and their predators. A bit of tenacity also is required to make sure the crop maintains an advantage over its pesty competitors. Some crop failures are inevitable, but as gardening skill levels increase, successes become more frequent and yields become more abundant. The transition from a beginning gardener to a master gardener requires a lot of learning and experience, but it does not take a lot of skill to get started.

The scale of horticultural production or size of garden is also quite flexible and accommodating of the variety of spaces available in urban settings. A small urban balcony or patio will accommodate a few tomato plants and a small herb garden. Kitchen gardens can replace existing lawn space or flower beds and flowering fruit trees can replace flowering inedible ornamentals. Vacant lots can be turned into community gardens, where neighbors can not only share tools and equipment but also can share skills, gardening tips, and experiences. Community gardens also allow participants to share nature's bounty, providing a wider variety of produce than individual growers have the skills or space to produce. Larger spaces, such as vacant lots, also can be utilized as market gardens, where skilled gardeners produce a variety of horticultural crops for sale to their neighborhood farm stands or local farmers markets.

Larger lots in peri-urban areas allow market gardeners and farmers to use larger and more efficient power equipment. Farmers with more land can produce more extensive field crops, such as green beans, green peas, and sweet corn. Products from such operations typically are destined for local markets, and with further expansion, peri-urban farming could make a significant contribution to local food security. *Greenbelts* around cities provide opportunities for farming, as well as open spaces and scenic landscapes that are protected by government land use policies. The *Garden Cities* movement of earlier times appears to be reemerging on a grander scale with the concept of greenbelts around major cities, such as Toronto, Canada.[8] A logical economic use of such protected spaces would be farming operations similar to the truck farms surrounding cities of the past—with an emphasis on fresh vegetables and fruits. In such areas, the potential for both commercial crop and livestock production also provide opportunities for holistically managed, integrated, sustainable farming operations.

<u>The Role of Government in Urban Food Security.</u> An absence of government subsidies or public financial support of urban agriculture is not a requisite for its economic viability. Food security is a fundamental government responsibility and has been the political rationalization for virtually all U.S. farm and food policies. Governments

[8]Greenbelt Administration, "Farming in Ontario's Greenbelt; Opportunities Grow Here," October, 2013, http://www.greenbelt.ca/farming_in_ontario_s_greenbelt_possibility_grows_here.

simply allow people, as citizens and taxpayers, to make collective investments that benefit the "common good" rather than individual self-interests. Food security is not purely an individual matter. The right to life is a fundamental human right, and what could be more important to live than safe, nutritious food. A fundamental purpose of any government is to ensure the basic rights of the governed.

A sustainable agri-food system must be economically viable, but it is important to understand that economic sustainability does not suggest maximum economic efficiency. In fact, today's unsustainable agri-food system is the natural consequence of pursuing economic efficiency at the expense of ecological and social integrity. Economic value is individual, impersonal, and instrumental Ikerd (2012). The economy places no value on doing anything solely for the benefit of anyone else and certainly not for any of some future generation. The need for economic efficiency must be balanced with the need for ecological integrity and social responsibility—which can be accomplished only through effective governance.

The U.S. government has supported and continues to support the industrial approach to agriculture by providing generous financial subsidies as well as a highly preferential tax and regulatory environment. The initial justification was to promote a system of farming that would provide domestic food security by making agriculture more efficient and food more affordable. The failure of today's industrial agri-food system to provide food security—in terms of either accessibility or quality—is becoming apparent to a growing number of Americans. It now seems logical and reasonable that government farm and food programs—at national, state, city, and community levels—should shift to supporting sustainable alternatives to industrial agriculture.

Government provides a legitimate means of giving the "common interest" of people in urban areas priority over the individual economic self-interests of corporate investors in the agri-food systems. Governments can legitimately support community gardens that promote social equity and justice by supplementing the diets of those in urban areas abandoned and ignored by the market economy. Urban agriculture can make good food accessible to everyone in the neighborhood, rather than just those who can afford to buy organic and other sustainably produced foods in supermarkets and restaurants. Even if urban gardens cannot provide food self-sufficiency, they can create neighborhoods where people are committed to ensuring enough food as a basic human right—at least in their own neighborhoods.

Caring communities can make good food accessible and affordable to everyone without making food cheap. Families in low-income urban communities may spend 30 to 40% of their disposable incomes for food, while food claims less than 10% of the average American's income. However, less than 20% of total food spending actually pays for food. The rest, more 80%, goes to cover the cost of processing, packaging, transportation, advertising, preparation, and distribution—rather than the cost of producing agricultural commodities. This spending may make foods more convenient for consumers to secure and prepare. However, processing, packaging, and storage make food more costly as well as less nutritious. A family that spends 40% of their income for food could save more than one-third of family income by growing and preparing their own food. Their food would be fresher, more flavorful, and more nutritious—and would cost 80% less than they are currently paying for food. Urban families will not be able to grow and prepare all of their own food, even in the most favorable climates. But, each dollar's worth of food they can produce and prepare at home will save them something close to 80 cents of each dollar currently spent for food.

Knowledge of how to grow and prepare raw or minimally processed foods obviously are significant obstacles to providing urban food security through urban agriculture. However, these are obstacles that can be overcome through local, state, and federal government support for the caring communities that form around community gardens. Finding time to prepare food from scratch is another frequently mentioned obstacle. However, top chefs are proving that many foods, particularly fresh picked vegetables and fruit, require little preparation to create delicious meals. Also, family meal preparation can be a far more productive and socially constructive activity for family members than watching television, playing video games, or just hanging out on the street corner. Urban food production and home preparation can also help young people develop a sense of self-reliance and expose youth to intergenerational interactions that have been disrupted by industrialization. Urban agriculture alone obviously cannot solve the social problems of inner cities or even ensure urban food security. However, it can play a more significant role in both that is commonly assumed by economists and others who influence government programs meant to provide food security for people in urban communities.

Contribution of Urban Agriculture to Sustainability. A sustainable agriculture must be ecologically sound, socially responsible, and economically viable. Everything of use to people, including everything of economic value ultimately comes from the earth—air, water, minerals, soil, and energy. There is no alternative. The connections of agriculture and food production to the soil, water, air, or earth are just more direct and obvious. Beyond self-sufficiency, people must depend on other people to meet needs they cannot meet through direct relationships with nature. Economies simply provide impersonal means or markets through which people can meet their needs by working, selling, and buying, rather than sharing, trading, or bartering. Thus, the sustainability of today's complex modern economies depends on maintaining the integrity and productivity of the earth's natural ecosystems and civil societies.

The basic needs of both current and future generations can only be met by societies willing to balance their economic their individual self-interest with the greater good of society and humanity—giving the basic laws of natural ecology priority over their societal values and the needs of society priority over individual economic self-interests. Government is simply a means of allowing people to give the long-run common good of society and humanity priority over short-run individual preferences. These fundamental principles of sustainability are reflected in the multidimensional motivations for today's resurgence in urban agriculture.

Urban agriculture obviously cannot sustain urban economies, but it still can make significant contributions to urban employment and economic development. Underemployed, low-paid workers in inner cities can make up for the deficiency in the quantity and quality of their current employment by engaging in urban agriculture. Urban gardening can offset at least some of the lack of dignity in their current employment as well as the deficiency of income. Those who are unemployed may develop their farming skills by working in home gardens and community gardens to the point where they can assume paying jobs as organizers or managers of community gardens. Some may start their own full-time urban gardening enterprise. Some commercially successfully community supported agriculture organizations, or CSAs, have been developed using neighbors' backyards as their "farmland." Commercial CSAs, farm stands, farmers markets, home delivery and other urban food ventures may not be sufficient to revitalize the economies of inner cities, but they can contribute economically as well as socially.

Urban agriculture can also contribute to the ecological integrity of urban communities by cleaning up polluted vacant lots restoring life and health to city spaces. Urban farmers can turn parks and green spaces into community organic gardens and orchards, rather than chemical hungry lawns and imported ornamentals. By engaging individuals and entire neighborhoods in promoting production of nutritious, healthful, food—free to harmful chemical residues—communities may become more conscious of the multitude of environmental risks that threaten the health and well-being of those who live in inner cities. Urban gardeners can organize to improve neighborhood health and safety by combating urban blight in general—economically, social, and ecologically. When people become empowered by regaining a degree of control over an aspect of their well-being as important as food, they naturally grow toward taking ever greater control of their own destinies.

In fact, the greatest potential contribution of urban agriculture to sustainability is social, rather than economic or ecological. Up to now, the focus of the sustainable agri-food movement has been on ecological integrity and economic viability. The emphasis has been on preserving opportunities for the future rather than meeting the needs of all in the present. Sustainability advocates have been preoccupied with finding economically feasible means of protecting the natural environment and conserve nonrenewable resources while giving little attention of the failure of the current economic and political systems to meet the most basic food needs of many today. Urban agriculture is a highly visible and increasingly important response to this failure.

Those in the sustainability movement cannot expect people whose most basic economic needs are not being met today to sacrifice still more in order to ensure opportunity for those of the future that they do not have today. Sustainability is based on the implicit premise that those of the future have rights "equal" to those of people today—that we should leave opportunities for the future at least equal to those people enjoy today. Most American can easily afford the economic sacrifice of ensuring an even better future. However, many cannot. When people in inner cities lack access to enough safe, healthful foods to meet the needs of their own families, they cannot be expected to support sustainable agriculture initiatives that they feel would make their food even more expensive or less accessible. The first responsibility of sustainable agriculture is to meet the basic food needs of all in the present. Only then will the sustainability movement gain the credibility it needs to become a widely acceptable public issue.

Nowhere is the potential for a truly socially responsible agri-food more evident than in the urban agriculture movement of today. People in urban areas are beginning to understand that a market economy alone will never meet even the most basic food needs of those who lack the economic ability to compete with other uses of agricultural commodities. They are beginning to understand that simply making food cheaper is not the answer. The 50-plus year industrial agri-food system experiment has failed them. They are coming together to build caring communities around their common interest in ensuring enough safe, wholesome food to meet the nutritional needs of their families—all of their families. They are giving priority to the common interest of the community because the pursuit of individual economic self-interests has left them poor and hungry. They are finding ways to afford enough good food by rejecting the costly convenience of highly process, packaged, and prepared foods. They are growing and preparing their own food. But most important, they are empowering themselves and their communities to take control of their own destinies. The urban agriculture movement, at its best, exemplifies the ethics of social responsibility essential for agricultural sustainability.

References

Colasanti K, Hamm M (2013) Increased productivity, role in alleviating food insecurity possible. J Agri Food Syst Community Dev 3(2):15–16. https://doi.org/10.5304/jafscd.2013.032.011

Coleman-Jensen A, Gregory C, Singh A (2013) Household food security in the United States in 2013 (Economic Research Report No. [ERR-173]), p. 41. http://www.ers.usda.gov/media/1565415/err173.pdf

Evans TL, Miewald C (2013) Cultivating more than food: Where community gardens fit with what cities do. J Agri Food Syst Community Dev 3(2):19–21. https://doi.org/10.5304/jafscd.2013.032.013

Hallsworth A, Wong A (2013) Urban gardening: a valuable activity, but .... J Agri Food Syst Community Dev 3(2):13. http://dx.doi.org/10.5304/jafscd.2013.032.010

Howard E (1902) Garden cities of to-morrow. Swan Sonnenschein & Co, London. https://archive.org/stream/gardencitiestom00howagoog

Ikerd J (2008) Crisis, and opportunity—sustainability in American agriculture. University of Nebraska Press, Lincoln, NE

Ikerd J (2012) The essentials of economic sustainability. Kumarian Press, Bloomfield, CT

Lavid L (2013) Urban gardens: Part of a whole system approach. J Agri Food Syst Community Dev. https://doi.org/10.5304/jafscd.2013.032.012 Advance online publication

McLean T (2017) Michigan state university extension, urban agriculture trend continues in 2017, 26 Jan 2017. http://msue.anr.msu.edu/news/the_urban_agriculture_trend_continues_in_2017

Mumford L (1961) The city in history, its origins, its transformations, and its prospects. Houghton Mifflin, Harcourt, New York. https://books.google.com/books/about/The_City_in_History.html?id=q0NNgjY03DkC

Royte E (2017) *ensia*, “Urban agriculture is booming, but what does it really yield?” 27 April 2017. https://ensia.com/features/urban-agriculture-is-booming-but-what-does-it-really-yield/

Smit J, Nasr J, Ratta A (2001) Urban agriculture: food jobs and sustainable cities (Chapter 2, edition, published with permission from the United Nations Development Programme). http://www.jacsmit.com/book/Chap02.pdf

Weissman E (2013) No buts about it … the value of urban food production. J Agri Food Syst Community Dev 3(2):24. https://doi.org/10.5304/jafscd.2013.032.014

Wojtowicz R (2001), City as community: the life and vision of lewis mumford, vol 4, issue 1. Quest, Old Dominion University, Jan 2001. http://ww2.odu.edu/ao/instadv/quest/CityAsCommunity.html

Index

D. Nandwani (ed.), *Urban Horticulture*, Sustainable Development and Biodiversity 18, https://doi.org/10.1007/978-3-319-67017-1

The manufacturer's authorised representative in the EU is Springer Nature Customer Service Centre GmbH, Europaplatz 3, 69115 Heidelberg, Germany. If you have any concerns regarding our products, please contact ProductSafety@springernature.com

Printed and bound by CPI Group (UK) Ltd, Croydon, CR0 4YY
30/09/2026
02225587-0003